T0341109

Real Estate Crowdfunding

Real Estate Crowdfunding: An Insider's Guide to Investing Online introduces the reader to basic real estate investment concepts and then takes a deep dive into how to invest passively yet wisely in real estate syndications.

This book will teach the reader how to:

- invest in crowdfunded real estate syndicates
- understand key financial concepts used in the industry
- diversify their investment portfolios
- read between the lines of investment contracts
- maximize profit while minimizing losses

This book is a guide to the foundational financial concepts upon which all real estate projects are based and explains the language of real estate from an insider's perspective. It provides a road map of what to watch for and how to win at the game of passive real estate investing.

Adam Gower is a real estate investment and finance professional with over 30 years of experience and holds a Ph.D. in how syndicates mitigate risk (an historical perspective) from University College London. He is an internationally recognized expert in real estate crowdfunding and has taught thousands of students, both at university level and online, the secrets of successful real estate investing.

Real Estate Crowdfunding

An Insider's Guide to Investing Online

Adam Gower

Routledge
Taylor & Francis Group

LONDON AND NEW YORK

First published 2021
by Routledge
2 Park Square, Milton Park, Abingdon, Oxon OX14 4RN

and by Routledge
52 Vanderbilt Avenue, New York, NY 10017

Routledge is an imprint of the Taylor & Francis Group, an informa business

British Library Cataloguing-in-Publication Data
A catalogue record for this book is available from the British Library

Library of Congress Cataloging-in-Publication Data
Names: Gower, Adam, author.
Title: Real estate crowdfunding: an insider's guide to investing online / Adam Gower.
Description: First edition. | Boca Raton: CRC Press, 2020. |
Includes bibliographical references and index.
Identifiers: LCCN 2020028981 (print) | LCCN 2020028982 (ebook) |
ISBN 9780367428044 (hardback) | ISBN 9780367428068 (paperback) |
ISBN 9780367855239 (ebook)
Subjects: LCSH: Real estate investment. | Crowdsourcing.
Classification: LCC HD1382.5 .G68 2020 (print) |
LCC HD1382.5 (ebook) | DDC 332.63/24–dc23
LC record available at https://lccn.loc.gov/2020028981
LC ebook record available at https://lccn.loc.gov/2020028982

ISBN: 978-0-367-42804-4 (hbk)
ISBN: 978-0-367-42806-8 (pbk)
ISBN: 978-0-367-85523-9 (ebk)

Typeset in Bembo
by Newgen Publishing UK

Contents

1   Introduction

I'd like to start off by acknowledging that you are (probably) not here because you have a deep and abiding curiosity about the exciting and glamorous world of commercial real estate and real estate syndications. You're here because you are searching for ways to secure your financial future and to diversify your retirement savings, and you're looking for unbiased, verifiable, factually accurate information about how crowdfunded real estate syndications can help you grow and preserve your wealth.

I'm going to kick off our journey by explaining why you really shouldn't be intimidated by the scope of learning about crowdfunding, and real estate investment as a whole. The concepts and jargon that underpin the commercial real estate space may seem daunting at first, but with a bit of study and a dedicated mindset, you too can become a competent or even an exceptional real estate investor.

With that being said, I'm certainly not trying to minimize this highly technical skill or oversimplify the profession and what's involved. But I do want to break it down into some components that, once understood, can help you focus on a specific area within real estate and become proficient at investing in that area. Then you can diversify into other types of real estate from there.

Let's start by drilling down on why anyone would want to invest in real estate and the benefits specific to real estate investments. This will help to contextualize the details we'll get into further on in the book.

Lifestyles of the rich and famous

Commercial real estate has historically been the exclusive domain of wealthy individual and institutional investors. Taking a brief look at how they deploy their capital will shed some light on how best practices have evolved over the years. Among the largest investors in the United States are endowments, pension funds, and insurance companies. Educational endowments have substantial capital resources that have to be invested one way or another to create revenue streams for the future of their institutions. Pension funds collect contributions and insurance companies gather premiums, and they each pool these funds and invest them so that they can make payments when the policies mature, decades into the future.

One illuminating statistic: the 300 largest pension funds in the United States hold around $6 trillion in assets. It's hard to conceptualize such a huge amount, but given that the gross domestic product of the United States in 2016 was around $18 trillion, this puts that $6 trillion into perspective. The largest educational endowments in the country are held by Harvard University, with $36 billion; the University of Texas System,

$25 billion; Yale, $24 billion; Stanford University, $21 billion; and Princeton University, also $21 billion.[1] These are huge endowments and, in combination with pension funds, an average of 20 percent of the assets are invested in real estate.

High-net-worth families and individuals constitute another major investor class. On average, they hold 25 percent of their portfolios in real estate. Despite this high concentration of real estate investments by the country's leading institutions and high-net-worth individuals, the average investor only has 3 percent of their net worth invested in real estate.

That's the average investor, and while a lot of investors have much more than 3 percent of their net worth invested in real estate, a not-insignificant share have invested zero in this industry. Now that the regulations have changed, real estate syndications can be crowdfunded online, making them available to everybody. As an individual investor, you can now expand your portfolio and invest directly in real estate in the same way that endowment and pension funds and high-net-worth families and individuals have done for generations.

In total, there's roughly $15 trillion of commercial real estate in the United States, which makes it the next largest asset class after stocks and bonds. The majority of these assets are owned by "the privileged few"—that is, institutional investors and high-net-worth individuals and families.

The opportunity to invest in this way in real estate markets has never been available before to the vast majority of individual investors. There is a good chance that this is new territory for you and for many others who, like you, have never had the opportunity to invest in real estate. You've had the opportunity to invest in stocks and bonds before, and perhaps even small real estate deals, but none of that matches the new opportunities available as a result of the democratization of real estate investment due to crowdfunding.

In 2015, the first year that real estate crowdfunding was widely available, over $435 billion of commercial real estate has been traded. Only a tiny fraction of that went out to crowdfunding; the vast majority of real estate investment dollars came from the same institutions and high-net-worth individuals we've been discussing. This trend is slowly changing with the advent of crowdfunded syndications, as awareness of the opportunity to invest in real estate grows and is embraced by the public at large.

Here's a perspective I'd like to share about how crowdfunding has opened up real estate to those who would otherwise be completely closed out of this market—aka the eternal renter. There's an entire population of people here in the United States for whom homeownership has become almost impossible; for many, this is not down to any fault of their own, but rather the structural changes that have taken place in how we construct, maintain, and incentivize growth in housing stock, and the impact on home affordability.

Consequences of the affordability crisis

In recent years, home buying has become increasingly unaffordable. Prices have risen dramatically, putting homeownership beyond the reach of a considerable portion of the population. Lending standards have tightened, making it difficult to borrow. With higher prices, larger down payments are necessary, which means that people have to save more money before they are able to buy. Even if a mortgage is obtained, it is now more expensive than ever to own a home, because prices are so high and this offsets the historical benefits of low interest rates.

Two things have resulted from the rise in the true cost of owning a home. First, many people who want to own a home have been forced to rent. As a result, they are missing out on the opportunity to benefit from real estate price increases. If someone doesn't own a home, they won't benefit from rising real estate values. They are in a class of people in the United States that is excluded from one of the greatest wealth-building booms of our time—the boom in real estate.

However, now that they can invest in crowdfunded real estate deals, a new opportunity has emerged. Instead of using that deposit to invest in a home, especially when not enough of a deposit has been saved, one alternative is to use the money to invest in real estate syndications.

Of course, this isn't the same as investing in your own home. If you own your home, you live in it. You acquire benefit from living in the home. In fact, when I first got into real estate in San Diego in the early 1980s, one of the principals I worked for always advised: "Don't think of your own home as an investment. It's where you live. It's not an investment."

His point was that as long as you can afford a house, it should be your home. It's not an investment. If you're looking for an investment, you must look outside of your personal home, because the way you relate to an investment is very different from the place in which you raise your family and live your life. It's a different mentality. While that is undoubtedly a good point, historically, many Americans have built wealth via their homes—buying them and living in them for decades before retiring and using the equity they have accumulated for their retirement.

However, even with the housing crunch as bad as it is today, if you cannot afford the down payment on a home but you do have savings that you'd like to invest, you can, for the first time in generations, still invest in real estate—even if it's not your primary residence from which you derive additional utility. This new opportunity to invest in real estate through crowdfunded real estate syndications allows those who cannot afford a home to put their savings to work in real estate assets without feeling closed out, and it has the potential to capture the myriad benefits of the real estate boom.

To be clear, I'm not downplaying the role or importance of homeownership in the United States or suggesting that someone not buy a home. It is, after all, the American dream to own one's home. But investing in crowdfunded real estate provides a complementary opportunity if you already own a home and an alternative if you do not. Indeed, if you do own a home, finding other real estate opportunities allows you to diversify your portfolio, and it provides an alternative way of investing in real estate to build wealth and earn passive income.

Six reasons to invest in real estate

There are six primary reasons to invest in real estate:

- higher returns
- passive income
- wealth building
- tax benefits
- unparalleled leverage
- diversification

1. Higher returns than traditional stocks and bonds

First, let's look at real estate investment returns by considering interest rates and the way your money works for you. Compared to other investment options, real estate offers fantastic returns. In 2015, on average, real estate returns were above 12 percent, which was significantly higher than anything you could get in the bank that year.

In the years after the Great Recession of 2007–2008, the most common scenario that most depositors could hope for was sub-1 percent interest from their bank. If they were lucky, maybe that interest rate would break 1 percent, but not by much. When compared with the potential returns from a real estate deal, or any kind of investment that yields 12 percent, it does not take a 'rocket surgeon' to see the writing on the wall.

In recent years, returns from real estate assets have consistently outperformed the stock market. In the period from 2000 to 2019, annual real estate returns were 2 percent better, overall, than the S&P 500 Index. This doesn't necessarily apply in the very long term. If you look back to the early 1980s and the way that the Dow performed from that point, which we'll go over later in the book, a very different story emerges.

To be clear, the vast majority of investment advisors counsel against having all of your investments in one asset class, whether that be stocks, bonds, real estate, etc. A diverse portfolio is a healthy portfolio, and it is prudent to have a range of different types of investments—including some real estate as well as some stocks and bonds, cash, and whatever else you own—primarily as a hedge against failure or poor performance in any single asset class. I'll go into more detail regarding the concept of diversification a little bit later too.

2. Real estate assets offer consistent passive cash flow

The second reason to invest in real estate is cash flow. Real estate assets have the potential to supplement your regular income through passive income generated from properties, primarily from tenant rents. There are a couple of ways to approach this objective: you could buy real estate directly, or you could invest in real estate as a passive investor through a syndicated project.[2]

If you go the first route, you could buy a duplex or a fourplex or, if you have enough money, five plus apartment units, an office building, a strip center or some other kind of commercial real estate. You will put a lot of money down and collect rent from the asset. If you get your calculations right, the income you get from the project will be more than the cost of owning and operating it. Total rents should be, obviously, greater than the cost of paying your mortgage and taxes plus maintenance costs and all other expenses.

And then, of course, there's the time that has to be invested in direct ownership. The importance of your time cannot be understated—it is the one thing even Warren Buffett or Bill Gates can't buy. The time you spend dealing with a direct real estate investment can take away from other ways of generating an income. Property management at the level you need to make a profit, or at least break even on your monthly property-related expenses, can in itself be a full-time job.

The other option is to invest in crowdfunded real estate syndications. Sponsors of syndicated deals are effectively creating a new type of capital formation that has its own unique idiosyncrasies. For example, the way the preferred return is paid has, in many cases, changed with the advent of crowdfunding. In the past, sponsors[3] would buy an asset and raise money from investors who would be paid a preferred return, depending on the type of investment. If the asset was already producing income, investors might receive a share of the rents on an ongoing basis from the first month they invested.

In other cases—for example, in a development deal—investors would invest their capital, typically into deals requiring hundreds of thousands of dollars as minimum investments, and not expect any returns until the business plan had completed its full cycle. The plan might be to buy some land and build on it; or to buy a strip center, evict the tenants, renovate it, and re-lease it; or maybe buy an office building with a similar value add approach. Investors were used to not seeing returns until the business plan was complete, which might be years in the future. Only then would they start drawing income and see return of capital with a profit share.

Things are different now. When you invest in a crowdfunded real estate deal, some sponsors pay you preferred return immediately, even in development projects where no active rents are being collected.

This may, on the face of it, appear attractive for investors, but in reality, it places strain on a project, because the only way a developer can afford to pay a preferred return on a project that is not itself producing rents is to raise more capital to finance the project. Put another way, investors are paid back their own invested capital to cover the preferred returns developers are offering—the money has to come from somewhere. So in these scenarios, crowdfunding can give you a path to investing your money today to get income immediately, but in reality, you're being paid with your own money. These deals seem attractive until you get to the nitty-gritty.

This is very different from the way real estate syndicates used to be capitalized. In the past, investors would assume that they would have to wait to reap the rewards of their investments; this, in turn, allowed developers to realize their business plans before returning capital, in any form, to investors. In cases utilizing the 'traditional' formula for real estate syndicates, preferred return was accrued but not paid current. Investors still earned 'interest' on their investment, but they had to wait for it to be paid out. One of the major benefits of investing in real estate is that you can expect to earn passive income; but with a prudent sponsor, you may have to wait until payments are made, depending on the modus operandi of the sponsor and when the project turns cash flow positive.

3. Real estate is a phenomenal wealth-building vehicle

The third reason to invest in real estate is as a vehicle for building wealth. When you make an investment in real estate, not only do you receive a share of cash flow distributions, but also you can hope to see the money you invested in the deal grow. That's called 'capital appreciation,' and it is how wealth is built. The capital—the cash you invest in the project—should grow by some multiple as the value of a building increases. This profit is in addition to your share of the profits from operating the building and collecting rents.

Keep in mind, however, that there is the possibility, as with bonds, and stocks, that your capital could be reduced if the overall value of the property you invest in declines. Have no doubt, investing in real estate is a high-risk proposition. You can make a lot of money, but you can also lose everything.

4. Real estate is among the most tax-advantaged asset class

The fourth reason to consider investing in real estate is the tax benefits. Before discussing this in depth, please keep in mind that I maintain one fundamental rule when it comes to anything concerning tax and accounting. It is the AAA rule: Ask An Accountant. I'm going to share some basic ideas with you, but I do not profess to be an accountant. In fact, throughout my career, anything with numbers on it gets sent to my accountant and I let her deal with it. She tells me where to sign and how much I owe, I write the check to the IRS, and I'm done.

That said, the accounting concept of depreciation presumes that buildings decay over time, and tax authorities provide tax breaks to compensate for this. The accounting principle of depreciation allows the real estate investor to deduct a portion of the value of the real estate they've invested in from other passive income they receive. This defers payment of taxes and can reduce long-term tax obligations as well.

Another prominent tax benefit that savvy real estate developers, especially those with longer-term perspectives, take advantage of is a section of the tax code that allows for all profits from the sale of a property to be shielded from tax through an exemption known as a '1031 exchange.' According to this exclusion, if a developer sells a property at a profit, provided they buy another property with similar characteristics to the one they sold, they are permitted to roll over any accumulated profit (capital gain) into the new building. Ultimately, tax is owed—there is no escape from that—but the burden of paying those taxes can, theoretically, be deferred indefinitely.

5. Real estate offers investors unparalleled leverage

The fifth benefit of investing in real estate is that developers are able to leverage the investments you make with them, which makes your money work even harder. I discuss the concept of leverage and what it means in Chapter 8 on the eight financial keys, but the Cliffs Notes explanation of the principle is that by taking low-cost debt from a bank, your investment returns have the potential to be amplified.

Keep in mind that while leverage is generally beneficial, too much of it can be bad. As Max Sharkansky, Managing Partner at Trion Properties, likes to say, "real estate doesn't kill deals—debt does." Banks are federally regulated institutions, which means that if a developer stops making payments on a loan they have taken out, the bank will have no choice but to foreclose on the property, taking ownership away from a developer and transferring it to someone else.

Typically, the only speculative investment you can borrow money for is real estate. The beauty of investing in a crowdfunded real estate syndicate is that the developer will take on the bulk of the risk.[4] Most banks will expect some form of personal guarantee on the loans they make. As an investor in a crowdfunded deal, you are not going to be liable for repaying the loan to the bank; this will be the developer's responsibility. Your liability is strictly limited to the capital you put into a deal, but you still get to enjoy the benefits of the leverage the developer brings to the project.

6. Real estate offers excellent diversification possibilities

The sixth reason to invest in crowdfunded real estate deals is diversification. If you buy a single asset and own it directly, you're going to have to invest a significant down payment, so you're going to have a heavy weighting of your savings in one asset.

This is very similar to the way it used to be in real estate equity investment, in the good old days before crowdfunding. If you had a high net worth or were a member of the right country club and were invited by a friend or somebody you knew to invest in a great real estate deal, as a minimum, you would probably have been asked to invest hundreds of thousands of dollars into a single deal.

One of the beauties of modern real estate crowdfunding investment is that minimum investments are far lower, and you can split your investments into small shares in individual properties and spread them across multiple properties. Put another way, you can diversify your investments to mitigate risk. For example, on the Small Change website, which is a regulated funding portal (more on portals in Chapter 2), you can invest as little as $500 in a single project. Some sites, like GROUNDFLOOR, offer minimum investments as low as $10.

Consequently, if you've got $10,000 you want to invest, you could invest in 20 different deals—that is, if you can find 20 deals that allow you to invest only $500. Of course, if you have $100,000 you want to invest and you want to put $5,000 into 20 deals, you're looking at a reasonably decent range of opportunities. On real estate crowdfunding websites like the industry-leading CrowdStreet or Fundrise, for example, they aggregate deals for you in funds that offer similar diversification benefits.

Of course, diversification also applies to investing in real estate in addition to stocks and bonds. But now, thanks to crowdfunding, diversification within real estate itself is possible; you have the opportunity to spread risk by investing across a large range of real estate deals, allocating a small amount to each one.

Those are the six main reasons to invest in real estate. You may have been thinking of at least one of these reasons when you chose to read this book, and as we continue, the nuances of each will become increasingly apparent.

The real estate moguls of our time

When you drill down further to the fundamental basics of real estate investing, there are only two things you can invest in—debt or equity. You can lend money to a real estate deal in the same way a bank might lend capital. In that case, you'd be providing debt to the borrower; or you could invest for profit, in equity.

The distinction is that if you invest in a deal where you provide debt to the borrower, where the developer is borrowing money from you, typically, you'll only get an interest rate in return for that debt. If you invest in equity, you'll get a preferred return plus some share of the profit from the deal.

So when lending in the form of debt, your position is a lower-risk scenario, in which you solely earn a flat interest rate, compared to investing in equity in return for both interest and a share of the profit. In the latter case, you're going to have a higher risk profile for the investment you're making, because lenders are the first to be paid out of any net operating income (NOI) from rents or sale of the property.

Fortunes have been made in these two different ways of investing in real estate, either as a lender or as a straight-up real estate equity investor. I had a conversation with Lew Feldman, Chief Strategy Officer for KBS Direct, for one of my podcasts. He talked about the superstars of real estate. He and I have been involved in real estate our entire careers, so we know these names. They are the A-list of the real estate industry, yet these folks are largely unknown outside of real estate circles.[5]

As your experience in real estate investing expands, you will start to come across the names of these leaders and their companies, and as you start to get more involved, it might be worth getting to know who some of these personalities are. I'll run through a handful of them now.

Our first real estate investment superstar is Donald Bren. He's number 30 on "The Forbes 400" list of the top wealthiest people in America. His net worth is over $15 billion, and he is Chairman of the Irvine Company. If you're in Southern California, you almost certainly have come across some of Donald Bren's many assets—he owns 110 million square feet of real estate in Southern California. That is a lot of real estate: 500 office buildings and 40 shopping centers. He's a real superstar. As the wealthiest real estate investor in America, he is, of course, number one on my list here today.

The next in line is Related Companies Chairman, Stephen Ross. Related Companies is very well-known in the real estate industry and one of the companies you'll likely come across often. Ross is worth $6.6 billion, which is significantly less than Donald Bren but not exactly pocket change. He's got some considerable developments in New York, including Hudson Yards, which is the largest private real estate project in the United States.

We've discussed the rarified nature of the upper echelons of the real estate business, but this 'old boys' club' has also led to a situation where few women have risen through the ranks. Among the first women to break the glass ceiling in the real estate investment industry was Barbara Corcoran. She is perhaps the most famous female real estate entrepreneur on the planet, largely due to her starring role on CNBC's investment-oriented reality TV show *Shark Tank*. While working as a waitress in New York, she hooked up with a real estate company, learned the trade, and eventually went into business for herself. She founded the Corcoran Group with her then boyfriend with a small loan of $1,000, which she grew and grew into a business worth tens of millions of dollars.

Moving on to number four, Richard LeFrak, the scion of a real estate dynasty. One of the best ways of making money in real estate is by inheriting it—something I've always dreamed of but, alas, not something likely to happen, in this life at least. Nevertheless, it is a phenomenal way of coming into real estate, and Richard LeFrak is one of the great beneficiaries of that process. That said, he's also been hugely successful on his own. His businesses, or his family businesses, own more than 16 million square feet of real estate in the United States. His net worth? A little over $6 billion.

Next on our list is Ted Lerner. You're going to love this guy. Apocryphal tales claim he borrowed $250 from his wife to start in real estate, and now he owns 20 million square feet of commercial and retail space and hotels as well as 7,000 apartments. He's worth $5.4 billion. If you invest $250 in a crowdfunded real estate deal as a result of reading this book and end up being worth $5.4 billion, please do remember me.

The sixth and final real estate dynamo, Donald Trump, needs no introduction. To be clear, I should mention that this book is in no way, shape, or form designed to be political. I stay away from politics as much as I possibly can. I bring up Donald Trump because he is, of course, probably the best-known real estate investor in the United States. 'Huge.' He is a tremendously wealthy real estate investor and 'bigly' successful. He has numerous trophy assets around the country—typically, large Class A office buildings, hotels, golf courses, etc.

Again, the kinds of assets these men and women own and operate are assets that, up until now, you have not been able to invest in—simple as that. Thanks to the relaxation on general solicitation and the change in regulations that allow real estate crowdfunding to exist, for the first time in generations, you have the opportunity to invest in trophy assets. You too can now invest in the kind of real estate that has helped make the rich richer, in precisely the same way as the real estate moguls of the United States always have done, along with their friends, relatives, and associates.

Those are some of the superstars of real estate—people you can look to for inspiration and whose footsteps you are now able to step into as you start to look at real estate investment opportunities online. You can now invest in real estate not only just like these folks but, in some cases, alongside them; maybe not these specific individuals, but others just like them who have become extremely wealthy real estate investors and developers. It goes without saying that successful real estate moguls like those listed above are incredibly professional and adept at what they do. They have lifetimes of experience, and in some cases they have also inherited expertise and experience from their family. They were born into this life.

Remember that real estate investing is incredibly complex and must be taken seriously. After all, it is your savings and future on the line. There is real risk involved with real estate investment. It's vital as you move forward through this book and start looking at investing in projects that you don't invest anything you cannot afford to lose. In this book, I'm going to break real estate down for you into its most basic elements so you can better understand this incredibly complex process and be able to make well-informed investment decisions.

The critical path rule

There's one fundamental rule of real estate—the critical path rule. This applies equally to the investment and development facets of the commercial real estate business. The critical path must be followed to be able to develop any kind of real estate, whether that is a multifamily apartment building, retail shopping center, or office park. To get from A to B when constructing or renovating, real estate requires a series of steps that must occur in a precise, chronological order or otherwise risk going over budget or taking longer to complete. This can introduce unwanted and profit-draining variables to your investment.

This same rule applies to the way you evaluate, underwrite, and look at crowdfunded real estate deals when considering investing. The Critical Path Method is a fundamental real estate rule that helps you minimize the risk of missing something when evaluating a deal.

Think of the critical path in terms of building a house. If you break it down one step at a time, the first thing you've got to do is buy land. Once you buy some land, the next step is to build a foundation. After you've created your foundation, you're going to construct a frame. Following that, you're going to put a roof on top of it, and the steps go on and on until you have a fully built home. What I'm getting at is that you should think of the critical path in the same manner—a step-by-step process that ends with a complete, purpose-built structure, only instead of sheltering people, that structure is sheltering and growing your wealth.

The critical path in real estate development exists whether you see it or not. And each one of the different points along this critical path has its own critical path. Just as when you look at a map of the United States, there may be one set of directions for a drive from California to New York, but there are also individual directions within that overarching route—the steps you take to get from California to Nevada, Nevada to Utah, and so on. For a more real-estate focused analogy, let's look at the critical path for building the foundation of a physical construction project.

The first thing we have to do is survey the site. Then we grade the land. Grading land means to flatten it (basically), dig up some dirt, and clean out the vegetative debris. Then, we put the dirt back and compact it to a certain density. Next, after the grading, we survey again to make sure everything is as planned. Then we review the plans and mark the site according to where the foundation points need to go, and we indicate where the caissons and grade beams are going and where we are going to pour the foundation cement.

After we mark the site, we excavate. We dig trenches for the concrete and then build and install the steel cages that go inside the concrete. We pour the concrete, wait for it to harden, take some core samples, and make sure those come back okay. At the end of the day, we have a foundation.

The critical path for building a foundation is only one of the line items on the larger critical path for building a home or project. Each line item on the major critical path to developing something has its own sub-critical path. If you think about each stage of the real estate process as being one step on a critical path, rather than thinking about the whole thing, you'll be able to break down the process into steps that are easier to comprehend. And that is exactly how this book is laid out—to make things as easy as possible for you—because you have a tough, but potentially lucrative road ahead.

Don't worry—it's not that difficult

On the face of it, real estate seems incredibly complicated and multilayered, an almost insurmountable obstacle for a newbie investor. However, as you'll find in the pages to come, when you break it down to its component parts, it's really not that difficult to grasp and, ultimately, to master.

There are, for example, only five types of real estate and only four development strategies you can employ across these types, and only eight phases of development. And guess what? To be an effective real estate investor, you don't even need to understand all of these. You only need to specialize, at least at the beginning, in one type of real estate and pick one strategy for investing. Plus, there's only one aspect of the phases of the development cycle you need to worry about—the financial underwriting.

Finally, to unlock every financial concept across the entire industry, there are only eight keys to understanding any deal you look at. These are the eight financial keys that open all the doors to real estate investment, and you are going to go through each and every one of them in this book.

The one big secret to success in real estate investing is that there really is no secret. As an investor, you just need to make sure the sponsor has done all the hard work, has done good research, has been diligent, and has employed a systematic analysis of every aspect of the project in which you are planning to invest.

In short, you need to make sure he's done his job, because you are just *investing* in real estate; you're not *developing*. You don't actually have to execute on a development plan. You need to make sure the developer knows what he is doing and that he's covered all his bases, and this book is going to show you exactly how to do that and what to watch for.

Notes

1   These fund sizes are accurate at the time of writing.
2   See Chapter 8 on contracts for a deeper analysis of the concept of 'passive' investing.
3   The term 'sponsor' is often used in real estate synonymously with 'developer' and 'operator.' These all cover the same concept. Banks and lenders call this the 'borrower.' Whoever's raising the money—the developer, sponsor, operator—they're all the same.

4 To be clear, this doesn't apply to acting as a passive investor in a deal; only as a sponsor in a deal who takes on debt as part of leveraging your, unlevered, investment.

5 You can hear the conversation I had with Lew here: https://gowercrowd.com/podcast/223-lew-feldman-ceo-heritage-capital-ventures

2 Real estate crowdfunding background

This chapter offers an interdisciplinary perspective on how to invest in real estate syndications through crowdfunding, a review of the origins and background of real estate crowdfunding, and examples of crowdfunding marketplaces.

The need for capital

When a sponsor finds a real estate deal that they want to build, they establish a budget for their project and then set about financing the project. At its simplest level, there are two types of capital developers need to finance real estate projects: debt and equity.

Debt is usually provided to developable deals by banks and other regulated institutions. No matter where it comes from—whether from banks or insurance companies or even from private lenders—debt, in most cases, won't cover all the costs of a real estate project. You are probably familiar with this from your own personal experience. When you buy a house, you get a mortgage, and in most cases banks don't lend 100 percent of the purchase price. Instead, you must put up some money of your own in the form of a down payment, which in commercial real estate is called 'equity' or 'risk capital.'

Sponsors have always had a variety of ways to raise the equity they need for a project. They can use their own money, but this becomes rapidly depleted as deal size and volume increases. To continue growing a development company and investing in more deals, a developer's next port of call for capital is friends and family. After that, they tap into what is colloquially called 'country club' money. This includes asking wealthy people in their business or social circles to invest in a deal. For more experienced developers working on larger-scale transactions, pension funds, insurance companies, and endowments offer access to capital either directly or through intermediaries like private equity groups.

Crowdfunding is another way to source capital for investment. Instead of the developer raising money from people within their immediate private network, using crowdfunding, they can instead look to a far larger number of people, either known or unknown to them, to finance their deals. In previous years, the term most commonly associated with real estate crowdfunding was 'syndication,' and this is still the mot juste today, though the laws pertaining to how a deal is crowdfunded have changed in recent years, as you will learn in this chapter.

The history of real estate syndications goes back to the late 19th century and early 20th century. Some of the first syndications were used for post-Civil War railroad financing. They had creative structures that, in many cases, almost guaranteed success for the sponsor whether the deal itself made money or not.[1] There are similarities between those early structures and the syndications permitted now through crowdfunding because of the Jumpstart Our Business Startups Act of 2012, also referred to as the JOBS Act.

An historical overview

In the 1920s, there was a stock market frenzy that ultimately led to a collapse of the markets and precipitated the Great Depression. People bought stocks believing the market would never stop going up. Some borrowed money and went into debt in order to buy stocks. Day after day, values went up and many people built what they believed was incredible wealth. Until the market collapsed. When it dropped off a cliff, these same investors lost absolutely everything, and some found themselves in debt to boot.

The 1933 Securities Act put a lid on that sort of frenzied investment by restricting the sale of securities to the general public without heavily regulated processes and transparency. Going forward, selling to the general public meant observing all kinds of rules, regulations, and disclosures, and there was a requirement to register an offering with the government. Unlike before, you couldn't go to the general public, hype them up, raise money, and then watch the whole thing collapse. The 1933 Securities Act was designed to stop all of that through mandating what had to be disclosed to investors when they were solicited to buy securities of any kind.

The Act also provided for exemptions from these rules for private offerings. Specifically, if you asked your friends to invest, you didn't need to register with the government or the Securities and Exchange Commission (SEC) to raise money for your deal, nor were there any onerous disclosure or reporting requirements. As long as there was a preexisting relationship with an investor, it was permitted for a sponsor to solicit investment from them. In some of today's contracts, you will likely still see references to this Act.[2]

Separating the public world and the private world

The 1933 Securities Act created a sharp dividing line between the public and the private world. You could raise money in the private world, but it had to stay private. Although you could go to the public sector, it was strictly regulated, highly controlled, expensive, and very time-consuming. On the other hand, using the exemptions from the Act meant raising money from the private sector was significantly less regulated, less controlled, and had almost no cost.

With it becoming increasingly costly, time-consuming, and burdensome to take projects to the public markets, real estate developers found themselves restricted to raising capital from investors in their immediate networks. Interestingly, and as with the 2012 JOBS Act, the 1933 Securities Act was never actually intended for real estate. It was designed to restrict unfettered liquidity in capital formation in the stock market and was geared to businesses. It was a completely unintended consequence that the real estate industry was also impacted.

The knock-on effect of the 1933 Securities Act on real estate was that over time, real estate developers found themselves raising money from closed, private groups that had little or no interconnection with other groups across the country, negotiating deals behind closed doors in increasingly exclusive groups. This closed-door approach to financing real estate became deeply ingrained in the way real estate works in the United States. Yet, as we'll see next, the 2012 JOBS Act changed everything.

The 2012 JOBS Act

In 2012, the JOBS Act brought major changes. As discussed, prior to the JOBS Act, raising money for a real estate deal required that a developer had to demonstrate a preexisting relationship with anyone they approached for investment. The JOBS Act is profoundly important in the history of real estate finance because it eliminated that requirement and opened the doors to general solicitation, which in turn meant developers could, for the first time in 79 years, advertise openly.

Importantly, by allowing open solicitation and advertising, for the first time ever, real estate developers seeking equity capital could use the Internet to publicize their deals. As you know, when the Internet comes to an industry, it changes everything. Some describe this as having a transformational effect on real estate finance; others call it revolutionary. Think of all the industries that have been upended because of the Internet. Just like those other industries, the real estate industry, previously a very secretive, closed shop industry, is now being disrupted.

Evolution of the accredited investor standard

As with all acts of Congress, those who write the rules and regulations that become the laws we must all abide by do their best to ensure that all contingencies are considered in an attempt to eliminate question and doubt about the intention of the laws that are promulgated. Seldom if ever, however, are resulting laws entirely clear from the outset, and adjustments are made over time once clarity is demanded by those using the new regulations.

While the laws pertaining to the 1933 Securities Act continued to evolve over time, there was one noteworthy component that remained particularly unclear. This pertained to who could be included in a developer's personal private network and came to be known as the 'accredited investor standard.' Even today, in order to invest in many crowdfunded real estate syndications, you must be an accredited investor.

The following is a description of the evolution of the standard and its current definition.

1953: *SEC v. Ralston Purina Co.*

It is useful here to look at the 1953 Supreme Court case *SEC v. Ralston Purina Co.* Ralston Purina had sold part of their stock to staff and employees, some of whom had very low incomes. Underpinning the exemptions to the 1933 Act was the idea that it should still protect those unable to 'fend for themselves' in an offering that did not go through the disclosure requirements of a public offering. The question before the court was whether an entity could solicit from investors who were not 'sophisticated' despite being, as in the Ralston Purina case, employees of the company and so individuals with whom the company had a preexisting relationship.

In this case, the court made a distinction between people who can and cannot fend for themselves, deciding that even with a preexisting relationship, people who are determined as being unable to fend for themselves when deciding how to invest their money must be protected.

The case defined who would and would not qualify even if a preexisting relationship existed, but left open the question of what it meant to be able to fend for yourself.

1974: SEC Rule 146

In 1974, in an attempt to clarify the questions left unanswered regarding who qualified for the exemptions afforded by the 1933 Act, and not fully clarified by the Supreme Court in 1953, the SEC adopted Rule 146 to address investors' level of sophistication. This rule defined sophistication as an investor's degree of knowledge, experience, and ability to bear economic risk. It narrowed the concept down but still didn't define what exactly it meant to fend for yourself or to be sophisticated.

1982: SEC Regulation D

In 1980, Congress authorized the SEC to define more precisely the concept of a sophisticated investor—one who could fend for themselves. From this, the definition of the accredited investor emerged, and in 1982 the SEC adopted Regulation D (or Reg D), which defines the accredited investor standard based on an income threshold.

Specifically, to be qualified for the exemptions to the 1933 Act, the SEC determined that accredited investors were those who had earned $200,000 of net income (or $300,000 for a married couple) in each of the prior two years and had a reasonable chance of earning the same in the current year, or those with a net worth of a million dollars or more. Meeting one of these qualifications meant that an investor was 'accredited,' and this became the de facto definition of being able to fend for yourself in a financial transaction.

Anyone not fitting this definition of an accredited investor was, therefore, non-accredited. The dividing line was simple: At a certain level of income or wealth, you were considered sophisticated and capable of fending for yourself. If not, then you were non-accredited.

So while the JOBS Act removed the preexisting relationship requirement and opened up the Internet for advertising deals (with some exceptions), investors still need to be able to fend for themselves—to be accredited—to qualify for exemption from disclosure requirements. You can now invest in any advertised deal without a preexisting relationship with the developer, but in most cases you must still demonstrate that you are accredited.

Accredited investors

As stated above, the accredited investor standard was defined in 1980 and promulgated in 1982, and up to time of writing the definition hadn't changed.[3] Yet the 1980s were a completely different era, as can be seen from the range of statistics provided in Table 2.1. For example, the Dow Jones high was 1,000 when the standard was first considered in 1980. Thirty-seven years later, the Dow had reached a high of 22,118. In 1980, if you put your money in the bank, you earned 12 percent interest; in 2017, bank deposits earned under 1 percent. To take out a mortgage in 2017 to buy a house, you would have expected to pay 4 percent interest or less. In 1980, you were charged 20 percent interest every year. And even though the minimum wage more than doubled between 1982 and 2017, based on inflation, the minimum wage (in 1996 dollars) actually fell.

Table 2.1 1980 versus 2017

	1980	2017
US population	227,225,000	324,985,539
Life expectancy	73.7	78.7
Dow Jones high	1,000	22,118
Dow Jones low	759	19,762
Federal spending	$591 billion	$3.27 trillion
Federal debt	$909 billion	$1.42 trillion
Inflation rate	13.5%	1.7%
Prime rate★	21.00%	4.25%
Unemployment rate	5.8%	4.3%
Cost of a new home	$76,400	$201,000
Median household income	$17,710	$55,775
Cost of a first class stamp	$0.15	$0.49
Cost of a gallon of gas	$1.25	$2.66
Minimum wage	$3.10	$7.25
Minimum wage in 1996 dollars	$5.90	$4.80

Note: ★The prime rate is the lowest rate offered, in aggregate, by banks to their lowest-risk borrowers.

In 1980, there were 550,000 accredited investors in the United States. In 1980, $200,000 a year in income was a lot of money, and if you had a net worth of a million dollars, you were very wealthy.

In 2017, there were 11 million accredited investors. Yet according to the SEC, only 300,000—less than 3 percent—had actually made investments. This meant there were 10.7 million people in the United States worth a million dollars or more each who had never invested in their capacity as accredited investors. For developers, this opened up an incredible new reservoir of opportunity, because it meant that there was nearly $11 trillion of untapped wealth that could be accessed for real estate finance for the first time.

Unlocking opportunity

For real estate sponsors (or anyone wanting to raise money), having access to this huge audience of investors via the Internet offers a new way to raise capital at a fraction of the time and cost. Before, raising money was traditionally a laborious task; sponsors would sit down over lunch or at a conference table and explain their background and track record and, only then, the deal. These days, developers using the Internet can employ autoresponder email sequences, hold webinars, record deal pitches and stories about their prior successes, and build relationships online with investors without ever actually meeting them. This extraordinary opportunity lets sponsors expand beyond existing relationships and find new investors; and by expanding their reach, sponsors have been able to reduce minimum investments from hundreds of thousands of dollars to tens of thousands or lower.

For investors too, the 2012 JOBS Act has unlocked opportunities that were previously inaccessible unless they were members of the exclusive social and business groups inhabited by real estate developers. What's more, because sponsors have reduced minimum investment hurdles, investors can place smaller amounts than ever before into individual deals, allowing for portfolio diversification that better suits the individual investor's risk-return profile.

The 2012 JOBS Act

The JOBS Act has seven sections, or titles. These are listed below, though the ones most relevant for real estate are Titles II, III, and IV.

- Title I: Reopening American Capital Markets to Emerging Growth Companies
- Title II: Access to Capital for Job Creators
- Title III: Crowdfunding
- Title IV: Small Company Capital Formation
- Title V: Private Company Flexibility and Growth
- Title VI: Capital Expansion
- Title VII: Outreach on Changes to the Law or Commission

JOBS Act changes at a glance

Titles II, III, and IV created the following rules for raising investment online.

- **Regulation D, Rule 506 (c) (Title II):** This allows for unlimited raises, is available only to accredited investors, and is relatively easy to implement, though verifying accreditation requires jumping through some hoops.
- **Regulation CF (Title III):** This is available to non-accredited investors also but must be processed through a funding portal (see the section below on Title III portals) and has limitations on amounts that can be raised and how much investors can invest; and
- **Regulation A+ (Title IV):** This is like a small-scale initial public offering (IPO) that allows anyone to invest, but with higher disclosure than Regs D or CF requirements, meaning that it takes longer to put in place and is generally more costly to launch.

The cost to launch these types of offerings start with the administrative costs of meeting legal requirements and getting the paperwork in place. Any competent securities attorney can put together the documents a sponsor needs. The big thing with all of these, however, is that having been given the right to market a deal online, this has now become necessary in order for the developer to be successful. There's little point going to the effort of putting together one of these offerings if no one ever hears about it. Anyone can put the paperwork together, build a web page and throw up a deal, but they still need sophisticated marketing processes to get the deal out and to draw people to it.

Developers today need to create digital marketing funnels to bring people in, nurture them through the investor relationship process, and eventually pitch their deal. They have to become digital marketing experts or else hire experts, like the team at GowerCrowd, to do it for them. In order to take full advantage of any of the regulations coming from the JOBS Act of 2012, developers need to become their own media companies selling investors on their deals.

Title II, Regulation D

Most commercial real estate crowdfunding syndications use Reg D to solicit investors. It is this regulation that governs the private placement exemptions discussed above. Under Reg D, there are no requirements to register the securities being sold with the SEC;

however, some other restrictions do apply, most notably the requirement to verify investor accreditation.

In general, developers using Reg D can make their offerings publicly known—though with some restrictions. On most of the major real estate crowdfunding marketplaces, sponsors use Reg D for their offerings. The regulation requires that some disclosures are made, but these are significantly less onerous than the requirements for any kind of public offering. As this regulation has been used historically in the decades prior to the relaxation on general solicitation restrictions, (an attorney would doubtless dispute the next characterization) the production of the legal documents for a Reg D deal has largely become commoditized, making it easier, quicker, and relatively inexpensive to launch.[4]

Title III, Regulation CF

Regulation CF (Regulation Crowdfunding, or Reg CF) is highly regulated. It came into law in May 2016 as a brand-new concept. With Reg CF, the Act permits a developer to advertise on the Internet, but with an upper limit on how much they can raise. Sponsors can't advertise deal terms on the Internet, but instead must create 'tombstones' that can announce the deal and point prospects to a crowdfunding portal where they can access all the detailed information they need to understand the deal.

Under the Act, Reg CF deals must be easy for anyone to understand, no matter their education level or degree of sophistication or experience in real estate investing. Anyone can invest in a Reg CF deal, whether accredited or not, and as long as they are over 18 years old. There are no limits on the number of investors, and there are detailed disclosures to file through the portal, but these aren't subject to regulatory approval like IPO disclosures.

The maximum that can be raised was initially set at $1 million in any 12 months, though this was subsequently revised upward to $1.07 million. There are frequent calls to increase the limit further to $5 million, though, at the time of writing, this has not been approved. There are limitations for non-accredited investors. If your income or net worth is under $107,000, then you can invest the greater of $2,200 or 5 percent of the lesser of your annual income or net worth. The portal manages or enforces those limitations and ensures compliance with these regulations.

Title III portals

Title III capital raising requires going through a portal, a quasi-government agency that is a regulated marketplace for trading securities. On a portal, issuers (those people raising money) offer and sell securities—shares in the company they own that is operating the real estate deal underpinning the investment. One of the portal's key jobs is ensuring that issuers comply with regulations: It protects the individual, unsophisticated investor (the non-accredited investor) who can't fend for themselves. Even if you are only investing $500, the government wants to ensure you are protected from scams and fraudsters.

Portals require expensive setup, are audited, and require employment of a compliance officer. They must register with the SEC and with FINRA (Financial Industry Regulatory Authority), and they are subject to FINRA examinations. They can charge fees to somebody raising money (flat fees, upfront fees, etc.) and may charge a percentage of the amount raised or take stock in a deal.

Overall, portals are highly regulated, and although they can sell data and advertising space, they can't offer investment advice. They also can't solicit for issuers. Instead, they must wait passively for people to come to them. They are permitted to be selective about who goes on their portal, so they can reject sponsor deals they don't like.

Furthermore, portals must have publicly accessible communications for investors. If a sponsor wants to raise money, they can't have private chats; all questions must be raised in a public chat room where the sponsor will provide answers, and that resource has to be available for everyone to see.

Title IV, Regulation A+

Title IV came into law in 2015. It permits investments from both accredited and non-accredited investors. Offerings can be for equity, debt, or convertible debt.

Commonly known as Reg A or Reg A+, Title IV has two tiers. Tier I, the lower tier, is less commonly used. It lets sponsors raise $20 million and has lower reporting requirements. Tier II, on the other hand, allows sponsors to raise up to $50 million in any single year. This is important because it's a tool that can be used in various ways. A Tier II offering is like taking a company public, but in a fraction of the time that it takes large companies to do this and at a fraction of the cost.

Reg A+ offerings function like IPOs but with a significant reduction in the required disclosures (an IPO is a massive undertaking with an enormous amount of disclosures). Some people call Reg A+ a 'mini IPO.' David Weild—who I write about in my book *Leaders of the Crowd*, which examines the origins of the JOBS Act—calls it 'IPO lite.'

Regulation A+ advantages

Reg A+ offers several advantages. Though similar to Reg CF in that non-accredited investors can also participate, raising capital using a Reg A+ structure increases the amount that can be raised from $1M to up to $50M. It also combines the financial muscle of larger, accredited investors with the political and public relations clout of neighborhood folk (non-accredited investors) who can come out to support the project during entitlement phases. With this second group, sponsors can offer investment opportunities to a new pool of potential investors who may themselves become future patrons of the property being developed.

Reg A+ also has its limitations. For starters, there are filing requirements and very detailed disclosure requirements. What's more, it is time-consuming and much more costly than other forms of offering. They can take a few months to launch and might cost $70,000 to $100,000 for the legal work alone. On top of that, there are annual reporting requirements.

As mentioned previously, there's little point launching an offering if nobody is going to hear about it. The biggest cost of launching a Reg A+ offering is the marketing cost. While marketing for a Reg D offering requires targeting accredited investors specifically, the marketing of a Reg A+ offering targets everybody. This inevitably makes it a far more costly proposition to advertise and market a Reg A+ offering effectively. Furthermore, as the minimum investment amounts in Reg A+ offerings are usually significantly lower than those in a Reg D offering, sometimes as low as a few tens of dollars, the eventual number of investors in a deal is typically vastly larger than in Reg D offerings.

Consequently, launching a successful Reg A+ offering requires attracting a significantly larger number of investors, and this also means a larger marketing budget.

Other Reg A limitations include the following.

- Fees charged in a project by a developer might be challenged by the SEC.
- Waiver of liability/fiduciary responsibility is frowned upon (as opposed to Title II, Reg D where almost anything goes).
- Filing has to be qualified (approved) by the SEC.
- Tier II unaccredited investors are limited to 10 percent of their annual income or net worth, whichever is greater. Though there is no limit for Tier I.
- Investors self-certify their status as accredited/unaccredited with no requirement on the issuer to reasonably verify this.

Yet even with all these limitations, the required work pales in comparison to a regular IPO, making it an attractive option for some developers and crowdfunding marketplaces.

Real estate crowdfunding websites

There are a number of ways to use the titles and regulations listed above for crowdfunding. Sponsors can raise money for individual deals, create fund vehicles for investors, or even recapitalize their holding companies. In this section, we'll review a few key crowdfunding marketplaces, but first we'll contextualize what they are and how they work.

Think about some of the websites where people buy and sell real estate that you may already know.

- **MLS (Multiple Listing Service):** This is a marketplace used for residential properties, commercial properties, land, condos, and apartments. Brokers list properties on an MLS so that other brokers and their agents can see them.
- **Zillow:** This is a marketplace for sales and renters which has become an Internet disruptor in the real estate industry.
- **LoopNet:** This is one of the key commercial real estate websites. This site has been around for a while and is the commercial real estate equivalent of the MLS.
- **Costar:** This is a commercial real estate website geared towards the larger-scale transactions in the commercial real estate world—downtown sky rises and other institutional-caliber properties. Costar is the industry-leading commercial real estate marketplace.

Importantly, all these sites are marketplaces for *whole buildings*. A seller will list a building or project for sale. In the case of the MLS, for example, they will list a single-family home for sale. A buyer will go to these marketplaces with the expectation that they will be purchasing an entire property.

In contrast, real estate crowdfunding marketplace websites are designed to sell partial interests in real estate deals, not whole deals. Deals offered on crowdfunding real estate platforms are not directly real estate deals. Sponsors form new entities—new companies—that they use specifically to operate individual real estate projects, and on the crowdfunding real estate platforms, they sell shares in those companies. Technically, they are selling securities in those companies, and this is why the laws that govern real estate transactions are subject to securities regulations and the SEC.

Investors going to these websites have no expectations of buying entire properties. They are looking instead for minority equity positions in the companies that own individual real estate projects.

Crowdfunding websites

In writing this book, I have been trying to keep all of the content evergreen so that no matter when you are reading this, it will be as relevant as it was on the day that I wrote it. This is not an easy task. Writing at the very beginnings of this brand-new industry, I cannot predict how things will change over time. I'm going to list a handful of prominent websites and marketplaces below. Some may no longer exist when you read this book, and other dominant players may emerge. For this reason, I will start the list with one of the most successful real estate crowdfunding sites to have emerged in the early years after the JOBS Act was passed, but which eventually was forced to close.

- **RealtyShares:** Under the helm of founder Nav Athwal, RealtyShares was one of the first marketplaces to emerge after the JOBS Act was passed. In the same way as the MLS or Costar aggregate listings for whole buildings on their websites, RealtyShares aggregated real estate crowdfunding projects on its website so that investors could invest relatively small amounts in individual projects. By the time the company folded at the end of 2018, they had attracted more than 120,000 registered users on their site, placed over $700 million of equity capital into over 1,000 projects, distributed $130 million back to investors, and raised some $80 million plus in venture capital.[5]
- **CrowdStreet:** In contrast to RealtyShares, CrowdStreet is one of the great success stories of the industry. Oriented to accredited investors only, in March of 2020, just six years after its founding, this platform was the first to announce that it had raised over $1 billion in equity, underscoring the potency of real estate crowdfunding. CrowdStreet is one of the marketplaces that offers investors access to individual transactions and third-party funds, as well as having creating their own blended portfolios of prescreened deals that investors can choose from to diversify their portfolios.
- **PeerStreet:** PeerStreet similarly offers low minimum investments in real estate deals. However, their focus is on the fix-and-flip sector. Specifically, PeerStreet focuses on the loans that hard money lenders make to fix-and-flippers. Founder Brew Johnson identifies best-of-class hard money lenders and has created a secondary market for their loans, buying their loans from them. The lender keeps the fees, but PeerStreet takes the paper, earning the interest owed on the loans. The fix-and-flip borrower then stops paying their original lender, the hard money lender, and instead makes mortgage payments to PeerStreet, which in turn crowdfunds the deal at a slightly lower yield to crowdfunding investors than they get on the original loan so that they can make money on the spread.[6]
- **Fundrise:** Founded by Ben Miller, Fundrise has elected to use Reg A+ to create real estate funds and to attract non-accredited as well as accredited investors. As of today, Fundrise has one of the most powerful marketing platforms in the industry, rumored to be raising anything from $1 million to $2 million a day for their funds. Their approach has been to create blended portfolios of real estate projects in the form of fund vehicles that are risk-adjusted so that investors can decide on their personal

appetite for risk and invest accordingly. Fundrise makes the equity investments in individual deals, and investors place their money in the funds that Fundrise manages. An investor can spread their risk exposure across multiple deals in a fund that is established to meet predetermined risk-return criteria.[7]

- **Small Change:** Run by Eve Picker, Small Change is one of the only active real estate Reg CF portals. Small Change also does what are called 'side-by-side offerings,' where a single project can raise money using Reg CF and Reg D simultaneously, thus including both accredited and non-accredited investors.[8]

- **KBS Direct:** As the third-largest owner of offices in the world, real estate powerhouse KBS buys skyscrapers and other major buildings. Using Reg D, KBS started raising $1 billion, $10,000 at a time, to invest in America's skyline. These are relatively safe core and core plus real estate investments, and the returns are smaller to account for the lower risk profile.[9] In my conversation with Chuck Schreiber (the 'S' in KBS), he told me that the average investor was investing around $80,000 in each transaction. KBS Direct subsequently increased the minimum investment requirement to $25,000. I asked Chuck why it was that a company such as KBS would raise money by using crowdfunding when they could raise as much money as they wanted from institutional investors. He told me that KBS wanted to provide access to investors who had never before had the opportunity to invest in premium downtown office building real estate, so they were trying to tap into this alternate capital source.[10] Furthermore, institutional investors often pull back during recessions at exactly the time when opportunistic investments appear in greater numbers but access to capital can be limited.

- **Feldman Equities:** This is a private company using real estate crowdfunding to raise capital and, like KBS Direct, is not a marketplace. Feldman Equities is a third-generation real estate company. They built the terminals at Chicago's O'Hare International Airport in the 1960s, and today they specialize in downtown office buildings in the Tampa area of Florida. A seasoned real estate sponsor with some 40 plus years of experience, Larry Feldman, the third-generation principal at Feldman Equities, understands that marketing has been the key to his success in real estate. Feldman Equities offers investors the opportunity, typically only available to institutional investors, to buy into major downtown skyrise office buildings. It is an example of a sponsor who has embraced real estate crowdfunding to go directly to investors.[11]

- **Trion Properties:** Trion Properties is another example of sponsors who have gone directly to the crowd to raise capital. Trion focuses on value add apartment investments on the west coast of the United States. Their first foray into crowdfunding was to list one of their properties on a major crowdfunding website. The experience was so successful that they decided to enhance their online presence, and they have built one of the most comprehensive websites in the industry to date, packed with resources for investors. Like Feldman Equities, and KBS Direct, Trion Properties only solicits investments for their own deals and, also in common with the other two, only from accredited investors.[12]

The marketplaces and sponsors listed above are indicative of the range of different types that exist. Of course, there are numerous other sponsors and marketplaces actively raising capital online, and the numbers are growing daily.

The potential for widespread change

Some people believe that crowdfunding real estate will change not only real estate financing but also the sort of real estate development that takes place, as the process becomes increasingly localized.

As we have seen, the way real estate financing evolved after the 1933 Securities Act was based on privilege, benefitting those who live in ivory towers and who will not be direct users of the properties they invest in. For these investors, who will not be affected by the nature of the eventual development, a standardized, pro forma approach to investment works best. Ask yourself, why is there a CVS or a McDonald's on every corner? Imagine you work for a multibillion-dollar pension fund with a billion dollars that has to be allocated to real estate. The only way you can do that effectively is by copying and pasting formulaic deals again and again. As a result, our society grows up with chains of similar stores everywhere, and all our towns start to look alike.

In contrast, with crowdfunding, changing the way real estate is financed means that there is potential to change the way America looks. Raising money from local, grassroots investors and putting local stores into retail centers promotes independent neighborhood businesses as opposed to large retail chains.

Let's say you're a developer with a local retail center project. You already have a loan of $1 million from the bank and you raised another $1 million using Reg CF. You have a $2 million deal with three or four shops—a coffee shop, an art gallery, or whatever—and a hundred local investors. By virtue of being part owners in the project, these local investors have a vested interest in seeing it succeed, so they are likely to come to the retail center, drink coffee, buy art, and do some shopping. In this way, more diverse developments, suiting local people's needs, are possible.

Real estate crowdfunding as game changer

To some people, crowdfunding real estate sounds like a scam. Even though the JOBS Act changed the way we do real estate, people still have this initial reaction to crowdfunding real estate, because it can be difficult to adopt new ways of doing things, especially in the high-risk world of commercial real estate investing. In the past, investors raised capital from private groups behind closed doors. Over time, these groups became wealthier and wealthier, were independent of each other and had little, if any, interconnection.

They fueled real estate development nationwide, but each group was typically concentrated in one geographical area and had limited access to real estate sponsors. As they became wealthier, the minimum investment requirement rose higher until real estate investing became a club only the wealthiest of Americans could join. Sponsors demanded minimum investments in the hundreds of thousands of dollars or more per deal, and the real estate landscape remained like this until the JOBS Act changed everything.

As I mentioned earlier, the JOBS Act wasn't intended to benefit real estate. When Congress went through the process of writing the JOBS Act, nobody, at any time, thought it would be for real estate; it was always intended to provide liquidity for small businesses. Yet real estate has been the biggest beneficiary of crowdfunding as enabled by the Act.

What's more, sponsors are getting wise. Since most people don't have hundreds of thousands of dollars to invest, sponsors are asking for investments as low as $500—in some cases even lower. Now hundreds of sponsors are raising capital for their deals either on marketplaces or, increasingly, by building their own digital marketing systems so that they can establish direct relationships with investors, disintermediating the marketplaces. Even with exceptions—you still have to be an accredited investor in many cases—you can invest in any deal that is advertised, whether you know the sponsor or not.

Notes

1 You can access a chapter from my earlier book, *Leaders of the Crowd*, that describes in detail how these early syndications worked at: https://leadersofthecrowd.com/

2 Check any real estate investment offering documents you've signed before; you'll likely see references to the 1933 Securities Act.

3 The accredited investor standard changed one more time in 2012 when the value of one's personal residence was excluded from the calculation of net worth. New regulations being considered seek to remove the use of income or wealth as a proxy for intelligence in part—as the current regulations effectively do—by including certain professional groups, such as those licensed as investment advisors and general and private securities offerings representatives, under the 'accredited investor' definition.

4 See Chapter 9, on contracts for a deep dive into what to look for in these kinds of offering documents.

5 RealtyShares was one of the early marketplace adopters of real estate crowdfunding regulations. It successfully raised tens of millions of dollars in venture capital, but ultimately collapsed, having grown too quickly due to the demands of its financiers. This is another classic example of how a business based on a solid foundation disintegrated, driven, as it was, by investor demands which were ultimately misaligned with the best interests of the company. Find out more about why RealtyShares failed at: https://gowercrowd.com/podcast/239-nav-athwal-realtyshares-special. And listen to the original conversation I had with Nav when RealtyShares was in its heyday here: https://gowercrowd.com/podcast/202-nav-athwal-ceo-founder-realtyshares

6 My podcast conversation with PeerStreet founder Brew Johnson can be found here: https://gowercrowd.com/podcast/206-brew-johnson-peerstreet

7 I've spoken to Ben Miller, the founder of Fundrise, a couple of times, once for my podcast—available at: https://gowercrowd.com/podcast/204-ben-miller-co-founder-ceo-fundrise—and again for my book *Leaders of the Crowd*—available at: https://leadersofthecrowd.com/

8 My podcast conversation with Eve Picker, founder of Small Change, can be found here: https://gowercrowd.com/podcast/209-eve-picker-smallchange She is also featured in *Leaders of the Crowd*.

9 See Chapter 5 for more on core and core plus investments.

10 Listen to Chuck Schreiber discuss why KBS Direct chose to crowdfund their deals here: https://gowercrowd.com/podcast/222-chuck-schreiber-ceo-co-founder-kbs

11 Larry Feldman, one of the early adopters among developers who chose to directly crowdfund their own deals, spoke to me. You can listen to this conversation at: https://gowercrowd.com/podcast/larry-feldman-ceo-feldman-equities

12 One of the most forward-thinking developers and among the first to directly set up their own crowdfunding systems, Max Sharkansky at Trion Properties shares his insights here: https://gowercrowd.com/podcast/219-max-sharkansky-managing-partner-trion-properties

3   Cycles

In this section, I am going to talk about the influence of cycles on the real estate market. Cycles can be dangerous. They are dangerous if you are riding them on the road, if you'll permit me the pun, and they are also dangerous when you think about the impact that an economic cycle can have on your real estate portfolio. They can cause the value of your portfolio to drop precipitously and, worse than that, put you in a position where you could actually lose everything.

I am going to go through an explanation of cycles and how you can protect yourself from them. As long as you are aware of what they are, you can watch out for them. It is very difficult—if not impossible, or even foolhardy—to attempt to time the market. Nevertheless, being aware of what cycles are is a critical step in ameliorating the risks of being caught unaware by a sudden downturn in the market.

Real estate cycles are driven primarily by the relationship between supply and demand, so let's start by taking a look at demand. Demographic trends are key drivers of demand. Population growth through increasing birth rate is one obvious example. Population aging is another relevant factor. This changes the composition of households, which in turn alters the nature of demand. With the aging of the population, we are seeing more and more households of only one or two people. This includes empty nesters, who are said to be 'overhoused.' Their children have grown up, but they still live in three- or four-bedroom houses. For various economic and social reasons, they stay put. This lowers demand in the sense that those folk already have housing and are not likely to move (and it also reduces supply because it simultaneously reduces housing turnover).

Immigration from overseas in some areas also drives demand by increasing the number of people in a geographical region. Domestic migration—people moving around the country—has the same effect. This can be people moving to Sun Belt areas like Florida or Phoenix for retirement, people who are motivated by looser state tax regimes, or younger people moving to high-tech areas like Silicon Valley or Seattle for employment.

Income, of course, is another very important demand factor. It does not matter how many people there are in a location—if they cannot afford to spend, there is no demand. On the flip side, areas with particularly high incomes will see high demand even if there is a relatively small population.

There are also employment trends unique to each area of the country; for example, the employment drivers of the automotive industry in the Midwest, the music and entertainment industries in Los Angeles, the tech industry in San Francisco. These trends drive demand because they attract people and increase the incomes of those who reside in those areas. The opposite can also be true. Take the Rust Belt, for example. Here you see very high unemployment, which has a negative impact on demand. Tied into this is income distribution, which drives demand because it has a direct influence on how much spending power people have.

Last on the list in relation to demand is transportation infrastructure. The first thing that comes to mind when I think of this is the building of new railroads. These are incredibly expensive and very long-term projects. When a new railroad—or subway system or any permanent transportation infrastructure—is built by a local municipality or state, the hubs around these new transportation stations create tremendous demand across a range of real estate types, including parking structures, retail outlets, hotels, offices, and residences.

When it comes to supply, the picture is slightly different. From a sponsor's perspective, when they look at a particular project, one key consideration is the inventory of existing spaces in the area. It is not good to build something that is in oversupply in the area. For example, you would not want to build new office buildings in a location that already has high office vacancy rates; nor would you want to build new apartments in a location with a large number of new construction projects in development.

The next thing to consider is how zoning and entitlements are structured in a particular area. Supply may be limited because a city or municipality simply does not want any more of a particular type of product in an area. This places a tremendous restriction on supply, and it is a good indication for developments with entitlement rights that there is going to be little competition from other developers.

Another way to examine supply is to look at permits in the pipeline at the local planning department. These records show if are any similar projects ahead of yours in terms of planning approval. Permits are a very good indicator of future supply in a given location.

Leasing terms and conditions required by local landlords are another indicator of a market's supply, and a sponsor can also measure supply by studying location characteristics and sales figures for competing centers. This includes looking at types of existing lease terms and current vacancy rates in that area. Sponsors must consider all of these factors to determine supply levels for a target project and decide whether a project is prudent and not oversupplied.

Overall, when considering a deal, you must be sure to examine how the sponsor has considered all the supply and demand factors. Replicating a sponsor's due diligence is difficult, but you do not really need to go that far. You just need to confirm that the sponsor has done their own thorough, detailed due diligence and review their conclusions and underlying analysis.

How supply and demand create market cycles

The relationship between supply and demand is unstable, and the way they interact is what creates the market cycle, as shown in Figure 3.1.

This diagram illustrates that there is no beginning or end to a market cycle; rather there is a continual cycle where one phase begets the next. But as the 'expansion' phase is at the top in the diagram, let's start with that. This represents a period of economic growth. After expansion, a contraction then leads to a recession in the market, followed by a period of recovery that slowly but surely leads, again, to expansion—and the cycle continues. It just keeps going, repeating over time in a somewhat regular fashion.

As it is very difficult to accurately predict the length of each phase in the cycle, it is important to remain cognizant of where we are at any moment in time and, more importantly, never to forget that there will always be a recessionary period at some point.

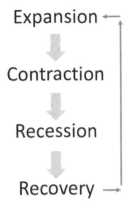

Figure 3.1 Market cycle

But why does the market not self-correct? Why do these four components—expansion, contraction, recession, and recovery—exist at all? Surely, when people demand something, there is supply, and when there is supply, pent up demand consumes that supply and the market finds equilibrium. That might be true with commodities, or stocks, where you go on to a market, ask to buy, find a seller, and trade immediately. Real estate is a very different beast. It takes a long time to trade, and that lag creates the boom and bust cycle.

Market inefficiency prevents the immediate self-correction of the market in real estate. There are various reasons for market inefficiency. For example, when a developer constructs a building and sets out to acquire tenants, there is a good chance some of those tenants may come from a different building in the market. In other words, the developer fills supply by creating a vacancy somewhere else, by cannibalizing existing buildings; in creating extra supply, the developer has taken demand and hence a tenant away from someone else.

Another factor contributing to market inefficiency is that tenants are more mobile than buildings. A tenant can decide to up and go overnight. Even with a contract in place, this can (in theory) be broken and the tenant can move somewhere else. But it takes considerably longer to construct buildings. This mobility is one factor that prevents the market from immediate self-correction. It's not that the market doesn't self-correct; what causes cycles is the fact that the market doesn't *immediately* self-correct.

Then, of course, there is imperfect knowledge in markets and of products. People simply do not know what is coming on the market. Developers cannot predict with accuracy what tenants are going to demand, and this imperfect knowledge creates inefficiency in the market.

The key thing to take away here, and the reason that markets do not correct immediately, is that development takes a long time and tends to happen in large increments, and supply does not fluctuate in sync with demand.

Development happens incrementally

In Figure 3.2, the *y*-axis (on the left) represents supply in square feet and the *x*-axis (along the bottom) is time. This chart illustrates how supply comes on stream in sudden bursts.

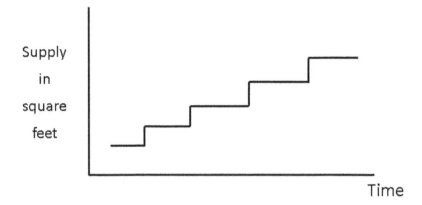

Figure 3.2 Supply over time

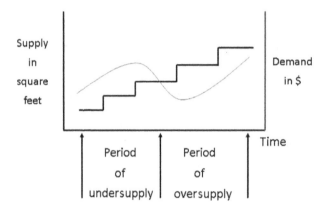

Figure 3.3 Undersupply and oversupply

Let's say you are going to construct a building that has 100,000 square feet. When a sponsor makes the decision to build, it takes a period of time to go through the entitlements and construction process. Then suddenly, overnight, the doors open and presto—there is another 100,000 square feet on the market. That is why we see this step effect in supply over time. It takes time for new supply to come on to the market, and when it does come on, it does so with a sudden leap.

In Figure 3.3, the previous chart is overlain with demand, indicated by the wavy line and given in dollars (see the vertical axis on the right).

There is a period of time when demand outstrips supply, which you can see on the left of the graph. This is a period of undersupply. Developers and sponsors might think: "Oh, look, there is a huge demand! We can start building!" At a certain point, because now they have this long lead time, new product suddenly start to come on to the market, and that means that supply is exceeding demand—a period of oversupply is shown to the right of the graph.

The impact of this in terms of rent is illustrated in Figure 3.4.

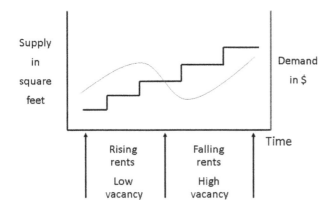

Figure 3.4 Rising and falling rents

During a period of excess demand—oversupply, where more people demand the little product available on the market—vacancies come down and rents go up. Vacancies get filled, and then landlords can increase rents. The opposite is true when supply exceeds demand. When this happens, you see rising vacancy rates and falling rents.

This applies across all asset types—offices versus apartments versus hospitality venues versus warehouses. Whatever it is, the same basic economic principles apply within each of the five different real estate types.

How oversupply happens

Professor Glenn Mueller, of Denver, describes the same idea using his "market cycle quadrants" diagram (Figure 3.5).

As you can see, phase one is recovery, where vacancies decline and no new construction occurs. Phase two is expansion. Here, vacancies continue to decline, during which time developers and sponsors decide it is a good time to build. Vacancies are decreasing; the economy is improving; they start building.

Point 11 on this particular chart indicates the entrance of phase three, or what Mueller calls hypersupply. Supply starts to exceed demand, so vacancy rates increase. Yet, because of the delay in starting construction and actual construction time, new construction and spaces continue coming on stream. So during the hypersupply phase, vacancies go up while supply continues to increase.

This leads to phase four, recession, where more and more product becomes available but the increase of supply starts to slow down or even stop. Developers and sponsors see vacancy rates increasing, and rents decreasing, so they stop making applications to build. Eventually, supply is choked off during the recession period. That, in turn, is consumed by new market demand, and vacancies start to decline. What happens here is that point 1 in the recession phase meets point 1 in the recovery phase, and the cycle begins again.

Professor Mueller's diagram appears again in Figure 3.6, but with an overlay demonstrating the impact on rent. In the recovery phase, you first see negative rental growth. In other words, rents continue to decline. Any rental growth that occurs after that is below the rate of inflation.

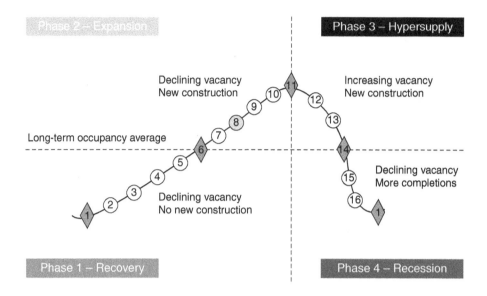

Figure 3.5 Market cycle quadrants
Source: Adapted from G.R. Mueller (1999) Real estate rental growth rates at different points in the physical market cycle. *Journal of Real Estate Research*, 18(1), 131–150.

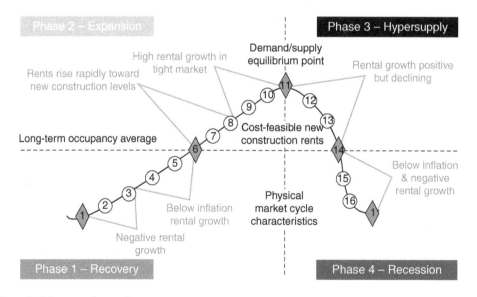

Figure 3.6 Rent cycle quadrants
Source: Adapted from G.R. Mueller (1999) Real estate rental growth rates at different points in the physical market cycle. *Journal of Real Estate Research*, 18(1), 131–150.

As we move into the expansion phase, rents begin to rise as vacancy rates decrease. As demand and supply reach a point of equilibrium, market rents start to peak. Mueller describes this as "high rent growth in a tight market."

When the hypersupply phase begins, rent growth slows. When the recession phase starts, rent growth dips to negative—that is, rents go down. Then the cycle starts all over again—back to phase one, recovery.

Cycles not waves

The phases in Mueller's diagram resemble waves more than a cycle. That might seem to be hairsplitting, but recoveries are generally very slow to take hold, and since the hypersupply and recession phases occur much quicker, a crash will be felt more strongly. In this way, markets behave like waves—you can have fun riding them but be ready for the crash. The market can suddenly turn on you, and it will not improve quickly. You need to be aware of this and give yourself tools to be prepared for what inevitably will come so that you are able to ride the inevitable downturn and thrive from its economic forces.

Some resources

To help you, I have recorded some podcasts with Professor Glenn Mueller. They are worth listening to if you want to know more about his thesis and the way that he views the market and analyzes cycles.[1]

I've also spoken to Charlie Nathanson at the Kellogg School of Management at Northwestern University.[2] Professor Nathanson discusses the idea that past performance is no guarantee of future results. We hear that all the time, especially when CEOs of public companies are giving their earnings reports. It is a disclaimer everywhere we go. It is so commonplace that we've become immune to it, and it tends to be ignored. What does this mean exactly? Well, the idea is that despite the warning that past performance is no guarantee of future results, people tend to hold on to the assumption that if prices went up last year, they will go up next year. Everybody is apt to think the same way; principals, developers, investors, and speculators.

The book *Winning Through Intimidation* by Robert J Ringer helps illustrates the point. This book was a bestseller in the 1970s, and some of the ideas still ring true. Ringer was a successful real estate broker in the 1970s. His book is not about how to intimidate; it's about *how not to be intimidated* in business. In other words, winning in spite of intimidation, or overcoming the ways that people try to intimidate you into caving in a negotiation. The other party wants you to think they're the big guy and you're little. If you get intimidated, then they win the negotiation.

In relation to developers, Ringer observes:

> I found that many apartment developers have a knack for getting their minds changed because of the fact that they are pathological builders, and that fact creates money problems. They cannot resist the temptation to keep building projects, and many times, they have no choice but to continue building, in order to keep enough cash circulating to pay for the last "I can finance out" deal, which, of course, ended up not financing out. A true pathological builder would not hesitate to construct a luxury hi-rise in the middle of the Sinai Desert if he could obtain the financing for it.[3]

I do not mean to be too critical of developers—I am one myself—but this is a truth that I have often seen. It is another driver of cycles—developers will attempt to build themselves out of a hole. If you get caught, as an investor, you can lose everything. You want to be really sure that your sponsor is building because the project makes economic sense and that their underwriting is oriented to recession resilience. A project has to be viable in and of itself, not because the developer is building to keep their team busy or to earn fees.

What Ringer's book is speaking to, of course, is a driver of supply that is very difficult to measure, and that is the developer's desire to keep developing and to stay in business just because that is all they know how to do. It is very difficult to say: "You know what? I'm going to stop the business of development. I'm going to shut everything down because I think the market is on a downturn."

Returning to Professor Charlie Nathanson's work, what he discovered about the housing market was that you have to be specific about what caused last year's price rises in order to predict what is coming up in the year to come.

You need to ask: What were last year's buyers thinking when they paid what they paid for a property? If they bought assuming that prices were going to rise, then their decision to buy was irrational. Their decision, in itself, was inflationary and disproportionate to market fundamentals. You want to see and measure whether last year's buyers were driven by market fundamentals or simply because they believed that this year's prices would be higher. A key indicator of the housing market is to look at the proportion of sales made to fix-and-flip buyers. This will tell you a lot about last years' purchase motivators, because fix-and-flippers clearly do not buy anything if they do not believe that the market is going to go up.

If you see a preponderance of fix-and-flippers in the market last year, it might be an indication that we are moving towards the hypersupply and recession phases. If you want to check on a particular market (maybe you want to buy a home), you would be well advised to figure this out. Brokers should be able to give you a sense of how many people were buying last year and how many of them were fix-and-flippers.

The other thing that Professor Nathanson found was that transactional volume alone is a precursor to where pricing is likely to go. If there is a sudden tailing off in volume, it may be the first sign that pricing may actually be on the way down.

Figure 3.7 shows a chart prepared by Professor Nathanson on the 2007–2008 recession that illustrates the point.

The dotted line is the transaction volume, or the number of sales (measured on the y-axis on the right). The solid line is the price index (measured on the y-axis on the left). After volume drops, prices continue to rise for a short period, plateau for a while, and then suddenly collapse. Think of that immortal character Wile E. Coyote when he runs off a cliff. There's the short period where he's running so frantically that he doesn't realize the ground beneath him has gone and that there's nothing holding him up. Once he realizes this, it's too late and he plummets. This is a lot like how the real estate market treats unwary investors—unless you are very careful, you don't realize how precarious the markets are until it's too late.

This is, however, just an indicator of where prices may be heading. You do not want to be buying at the peak of the market, and this indicator may help you avoid that.

Another interesting way of measuring where we are, potentially, in a cycle is the amount of inventory for sale, as measured by the months of supply left in the market. Based on this, we can consider how long it will take, if no more product comes on to the

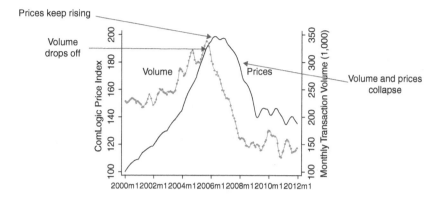

Figure 3.7 US housing market (2000–2012)
© Charlie Nathanson. Originally published in DeFusco et al. (2017) *Speculative Dynamics of Prices and Volume*. NBER Working Paper No. 23449. Available at: https://www.nber.org/permissions. html. Accessed 19 March 2020.

market, for the existing supply to be consumed by demand. In the early 1990s there was a peak of supply, as measured by months of supply in the market, equating to lower prices. Supply started to come down, dropping all the way through mid-2006. This was reflected in higher prices. Then, in early 2007, there was a rapid increase from around 3 months' supply of housing to, in some cases, over 12 months' supply, which was consistent with a collapse in house prices as supply exceeded demand and it took longer to sell homes.

Need to explore data in each market

Different cities experience differing degrees of reaction to economic slowdowns. They all suffer in some way, but some have greater exposure to downturns than others. Let's have a look at some markets and the different characteristics they have. Looking at the overall average US Home Price Index, we can compare how different markets reacted during the Great Recession of 2007–2008.

Figure 3.8 shows that, based on the Case-Shiller US Home Price Index, house prices peaked nationally in 2005–2006.

National prices reached a peak of 185 or so on the Case-Shiller index in 2005–2006, then with an overall decline of 27 percent, a low in 2011–2012. This indicates that it took six years, from 2006 all the way through to 2012, for the market to start to recover. That is a long time. However, by October 2017, the index was 6 percent higher than at the previous peak.

Now take a look at the Case-Shiller index for Boston over the same period (Figure 3.9). Here, the index came down by 18 percent and recovered 13 percent above the first peak.

Seattle shows a similar trend, though suffering a deeper recession with the index dropping by 32 percent after the first peak (Figure 3.10). By October 2017, the index was up 20 percent on the first peak. Compared to the national figures, this speaks to a significantly more volatile market. Would you want to be buying in Seattle in October 2017? It is important to be aware of these drivers and how different markets perform during downturns as you start to invest.

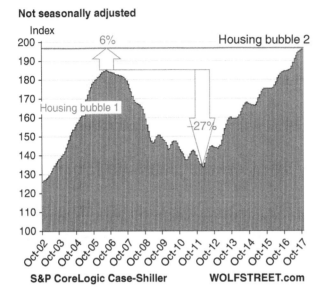

Figure 3.8 Case-Shiller US Home Price Index

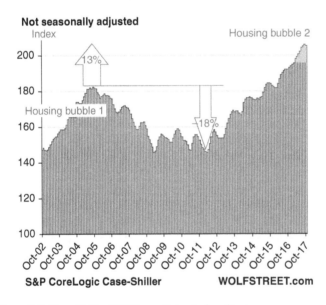

Figure 3.9 Case-Shiller US Home Price Index: Boston

Here are some more. Let us have a look at Dallas (Figure 3.11). It did not have a notable peak initially. The index was at 125 at its highest, but did not vary much. After the recession, the index rose by 43 percent above that prior peak.

What about Atlanta? Look at how the pattern is dramatically different in Figure 3.12. Atlanta came off its peak by 37 percent and was just barely squeaking above its prior peak in 2017.

Figure 3.10 Case-Shiller US Home Price Index: Seattle

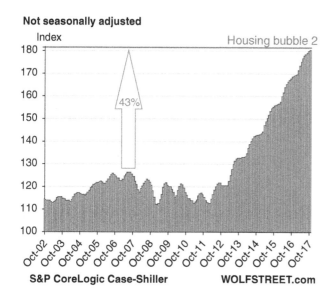

Figure 3.11 Case-Shiller US Home Price Index: Dallas

Finally, in Figure 3.13, you can see that by October 2017, Denver prices had risen to 44 percent above their prior peak.

Each of these markets experienced a different pattern, but they do show that it is important to recognize that cycles progress and prices do come down. When considering an investment, you need to look how the cycle will impact the local market.

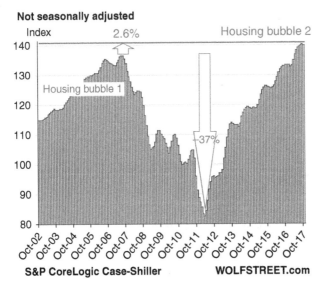

Figure 3.12 Case-Shiller US Home Price Index: Atlanta

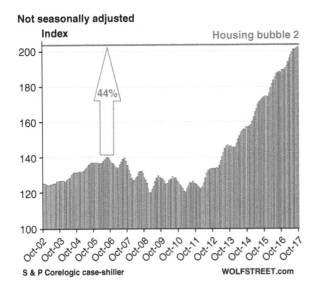

Figure 3.13 Case-Shiller US Home Price Index: Denver

Downside protection

The takeaway: Do not get caught in a cycle. Awareness is a great form of preparation. Just be aware the market is volatile. There are cycles, or waves. You will get caught in a cycle; there is no doubt about it. There is always a downturn at the end of an expansion phase, and you can lose everything that you have invested if you do not invest prudently and wisely. So do not invest anything you cannot afford to lose. It is the basic mantra for investing in real estate, but there are ways to help protect yourself against the worst impacts of a downturn, and some of these are listed below.

- **Buy good buildings.** Look for quality buildings in good locations with good tenants.
- **Consider the market in the area.** Determine how far a market dropped during the 2007–2008 recession for your asset type in the location you are investing, and underwrite to that level on the downside, layering only as much debt on your project as you can afford carry in that worst-case scenario.
- **Focus on long-term viability.** You may sacrifice returns, in the sense that you are not going to be looking at an 18 to 20 percent internal rate of return (IRR), but you are going to be better protected for the long run. It is more secure, and something you can count on for the future, if you invest for the long term (i.e. ten years and more).[4]
- **Explore various deals.** Look at a variety of deals. See how the sponsors are talking about the deals and how they are talking about the markets. Have a look at their risk summaries. You will find summaries of risks in the private placement memorandums. Have a good read of those—see what they say about the market and market cycles, and then see how they are actually addressing that. Do they acknowledge the possibility of a cyclical downturn? See what sponsors are talking about. What kind of language do they use to acknowledge the inevitability of cycles and how they are approaching it?

Notes

1 My podcast conversations with Glenn Mueller are available at: https://gowercrowd.com/podcast/010-real-estate-cycle-monitor-report-background
2 My talk with Charlie Nathanson is available at: https://gowercrowd.com/podcast/004-buying-into-a-bubble
3 Robert J. Ringer (1974) *Winning through Intimidation* (Fawcett Crest Books: Los Angeles).
4 Research and experience have taught me that ten years is the minimum optimal period to hold on to real estate to mitigate the risk of losing your investment. You can find out more in this article on my website: https://gowercrowd.com/learn/what-is-the-optimal-investment-period-for-real-estate

4   Five types of real estate

In this section, we talk about the five different types of real estate. While there are only five main types, each contains its own unique elements. For the most part, only a portion of what we discuss is currently available in the crowdfunding real estate investment space. In the near future, more of these asset types will adopt real estate crowdfunding to finance their deals, so it is well worth knowing what they are and how each asset is segmented.

Act like a developer—find a specialty

The best way to start out investing in commercial real estate is to pick the asset type that you feel most comfortable with and specialize in that. We will look at all the five asset types in this chapter, but keep in mind that developers specialize for good reason. There are very few developers who specialize in multiple real estate asset types. I can think of one—Chris Loeffler at Caliber in Phoenix. Soon after the passing the JOBS Act, this was one of the first companies to do a Reg A crowdfunding capital raise.[1] Caliber recapitalized their entire business and are a good example of a developer that does work in multiple asset types. However, this is more the exception than the rule.

Most developers specialize in one type of real estate and become proficient in that. Generally, they feel uncomfortable migrating away from what they see as their core competence.

In the same way, when you start to look at real estate investing, you might want to focus on one particular area first and get proficient in that. When you have built a foundation of competence and experience, you can then use this to gauge other options available to you and pivot into those as you become more expert.

The five main types of real estate are:

- office
- industrial
- retail
- residential
- hospitality

Office

We'll take a bit more of a dive into the different types of real estate, starting with offices. Although there are some existing industry standards, you will be hard-pressed to find consistency among industry specialists regarding what differentiates one type from another, but the following general categories of office building should hold you in good stead:

- low-rise: 1 to 6 floors
- mid-rise: 7 to 25 floors
- high-rise: 25 plus floors
- park or campus

There is also a distinction between metropolitan and suburban offices. As you can probably guess, metropolitan, or downtown, offices are more likely to be high-rise than those in suburban areas. Land in suburban areas is typically less expensive and at less of a premium, so it is unnecessary to construct costly high-rise buildings to capture economies of scale.

Within the office type are subsets based on the tenant. Some office buildings have a single tenant, while others have an anchor tenant with multiple tenants who take spaces with less square footage. The multiple-tenant building has no anchor tenant. These might have, say, seven or eight floors that are each subdivided, with each floor housing many smaller tenants.

Another differentiator in the office type involves the specialized use of an office building; for example, as a building with medical offices. This special use requires different development standards and different systems (floor plans, electrical requirements, etc.) to be invested in the property to accommodate the unique requirements of its tenants.

Parking is also a key differentiator in the office type. The parking provided with office buildings can be free or paid. This includes one-layer surface outdoor parking spread over multiple acres and structures specifically constructed and designated for parking. Interestingly, parking structures are now being designed with future reconfiguration or reuse in mind, as the need for significant parking capacity is likely to decrease with the rise in autonomous vehicles.[2]

Industrial

At the time of writing, despite being a huge asset type, there is little opportunity to invest in industrial assets via crowdfunding. Typically, industrial buildings are found on the outskirts of cities, primarily because the rents they command support relatively low land values. The rents that can be obtained from industrial buildings would not justify industrial real estate development in downtown areas, where land can be very expensive and in short supply.

Within the industrial asset type, you will see warehousing versus manufacturing. You may come across an operator who is trying to finance a warehouse strictly to be used for storage and distribution. Self-storage is considered a subset of the industrial asset type, although local zoning regulations sometimes classify such facilities as general commercial.

Next, you have the Flex/R&D industrial buildings, which are hybrid warehouse spaces for research and development. These kinds of building can be made up of 100 percent office space, though this is not the norm. The classic warehouse distribution facility commonly holds less than 15 percent office space, primarily just for administration. Manufacturing and assembly buildings typically hold less than 20 percent office space for administration.

Transportation is clearly a critical factor in this asset type. A downtown area is not an efficient location for an industrial building when it comes to receiving, storing, and distributing freight. The ideal location is one that is in close proximity to important transportation corridors, including accessible roads, railroads, airports, and/or seaports.

Retail

Retail is one of the most talked about asset types, primarily because of the idea that the Internet is eroding the demand for retail property. Certainly, it is the most dynamic type, meaning that it is changing more rapidly than the other types of real estate.

Retail includes single- and multi-tenant properties. The market is constantly segmenting and re-segmenting as new retail comes in and old retail goes out. In fact, there seems to be a need for continual repositioning of retail. The configuration of good retail development is quite an art. Retail developments are often prized by cities for their fiscal benefits, because they receive tax revenues from the commercial operations as increased consumer foot traffic is generally considered good from an economic perspective.

There are various types of retail centers. (Reminder: These definitions are not set in stone.) The strip center, a familiar variation, is usually no more than 30,000 square feet. Main Street is another subset, where commercial districts reside on both sides of a main thoroughfare.

Another type of retail center is the neighborhood center. These typically include big-box anchors, such as Target, Walmart, or a local grocery store, for example. Neighborhood centers typically range from 30,000 to 400,000 square feet in size. The anchor stores have smaller in-line tenants situated in between them.

Gaining in popularity is the lifestyle center, a destination property that combines entertainment, food and beverage, and retail establishments to draw consumers. You might visit a lifestyle center for dinner and a show, but there is an underlying expectation that while you are there, you will do some shopping too. As people become more accustomed to shopping online, these locations meet a variety of needs in one conveniently located distribution venue.

Finally, regional malls and outlet malls are among the largest of the retail centers. These colossal indoor retail facilities, which range in size from 400,000 to 800,000 plus square feet, often house several anchor tenants surrounded by a wide variety of retail and food and beverage establishments.

As I mentioned, retail is in a particular state of flux, regularly popping up in the headlines due to the impact of the Internet on shopping trends and habits, along with a plethora of myths and legends. If you are interested in learning more about the retail industry, the next section has some excellent resources for you.

Retail industry resources

The biggest retail industry organization is the International Council of Shopping Centers (ICSC), which hosts yearly trade shows. In collaboration with *The Wall Street Journal*, they created the Shopping for the Truth project, which addresses some of the big questions regarding the general decline of retail that have become fairly commonplace in the news and media.[3]

Another retail industry resource is a podcast that I did with Chad Syverson, a professor at The University of Chicago Booth School of Business. For some of his research, he partnered with Steven Levitt of *Freakonomics* fame. In the podcast, he discusses some predictions from his paper on retail trends over the next 25 to 30 years.[4]

It is very difficult, of course, for economists to predict what is going to happen tomorrow, let alone 25 to 30 years in the future, but he has some fun with the statistics and makes some predictions about the complete demise of retail and migration to e-commerce. Clearly, some of his thoughts conflict with what the ICSC would like to project. So from your perspective as a real estate investor, it is certainly worth a listen so you can contrast his opinions with the ICSC's more bullish outlook.

Residential

Now we move on to the asset type that I am personally most familiar with—residential real estate. Apartments or multifamily residential units are the most popular among investors. Certainly, on crowdfunding websites there is a preponderance of multifamily residential deals. Residential economics are the most intuitively understandable given that we all live, or have lived, in some sort of residence and experience it on a day-to-day basis.

There are several types of residential multifamily properties. One is the low-rise, or garden-style residence, including walk-ups with two to four floors, potentially spread across several acres. Mid-rises typically have five to nine floors, and high-rise properties contain ten or more floors. Apartments can be supply constrained because of zoning, development costs, and the tax environment. According to some, overzealous local jurisdictions also restrict supply—by making developments economically unfeasible—when they impose constraints on apartment developers to include, for example, affordable housing options.

At the higher end of the market, as the real estate cycle reaches its peak, land prices rise, forcing apartment developers to underwrite according to increasingly higher rents to make their projects pencil (i.e. make sound economic sense). What tends to happen as the cycle peaks is that the apartment units that are developed sit vacant, typically, at the higher end of the market, and this leads to a glut of luxury apartments and a shortage of affordable units.

In turn, this fuels local political pressure to balance housing options through legislation, ordinances, or regulations. This then creates an imbalance between supply and demand for apartments and can exacerbate the natural economic cycle.

The demand for apartments shifts according to a variety of factors, such as demography, employment, etc. (as discussed in Chapter 3). In the current environment, many people are living in their homes for longer, which means housing gets released to the market in smaller numbers, restricting supply.

Much of this has to do with an aging population. Many baby boomers are reluctant to move from their single-family residences, and this contributes to the supply of housing being restricted, particularly in states like California. This forces people into apartments and puts pressure on supply, which has a knock-on effect on rents, forcing them up to accommodate the increased demand against restricted supply.

Of course, one of the biggest drivers of residential demand is employment. With people tending to want to minimize commutes, there is increased demand for residential apartments around employment hubs like tech centers. A great example of this is in Milpitas, up in the Bay Area of Northern California, where major tech companies like Google are buying up every piece of commercial space that they can find. That, in turn, is driving up rents and demand for apartments in the area.

Then there is the trend for developing apartments in downtown cores, which, in some regards, is the definitive elimination of the commute, especially in major metropolitan areas. If you can live close to where you work, essentially you have no commute.

If you want to know more about this topic, you will find it listening to the podcast that I did with Chuck Schreiber of KBS Direct.[5] At time of writing, they were raising $1 billion using crowdfunding marketing methodologies. At least part of Chuck's investment strategy involves downtown core plus office building investments in close vicinity to residential developments in major metropolitan areas. That way, he captures the synergy I describe between apartments and workplaces in downtown areas.

A relatively new asset type within residential real estate that has only come about since the global financial crisis of 2007–2008 is bulk single-family homes for rent. During that downturn, a great many people lost their homes to foreclosure, and as a consequence, they were unable to buy a home again. Their properties came on the market and were bought by large institutions for the purpose of renting.

A last example is student housing. This used to largely involve the leasing of standard apartment building to students without any adaptation, but in recent years this asset type has become more specialized with floorplans and room design more specifically tailored to the way students use their residences—for example, with common lounges and kitchens situated on individual floors.

Senior housing

The final residential real estate category, senior housing, is one that I have worked on quite intensively in the last few months for a client. There are four types of senior housing. First, independent living is where residents move out of their homes and into an age-restricted apartment building that offers a full range of services on an à la carte basis, including food, medical services, and transportation. As the name implies, residents who live in these properties are not dependent on the services offered, but they do have access to them as they want or need them.

The next level of senior housing is assisted living, which offers a range of services including meals, laundry service, some healthcare, housekeeping, and transportation, as well as providing assistance to those who are aging in place. When it comes to transportation, this can include motor vehicle transport to and from shopping establishments or medical appointments, for example, or personal transport to and from different areas of the facility, such as the dining room.

Also on offer is memory care for those who require care and treatment for Alzheimer's disease or dementia. This is a very highly skilled service provided in senior housing.

The final type within the senior housing asset type is the nursing home, which is primarily geared toward short-term recovery from surgery or other short-term medical needs that require intensive therapy or skilled nursing care which cannot be performed at home.

Senior housing is a fascinating field to explore. With aging baby boomers entering the market, there is an enormous wave of demand heading towards the senior housing industry over the next few decades.

Hospitality

Moving on to hospitality, it is important to note that hotels are highly dependent on the economy. They do extraordinarily well when the economy is booming, and they crash hard when the economy goes into recession.

During the last downturn, when I worked for a bank, we actually took possession of a hotel for a week, before selling it to a buyer. It was quite challenging. How do you take ownership of a hotel? If it is an apartment building and you own it for a week, there is not much you have to do. With a hotel, you have to serve muffins in the morning and change bed linens. In some ways, it is similar to senior housing in that it is really real estate with a built-in operating business. You have to consider the skill of the operator to really know if they are going to be able to pull off a hotel deal successfully.

There are different markets for hotels, such as the conference/convention/business traveler market versus the leisure market. Within the different hotel markets are different hotel types. Full-service hotels offer on-site food and beverages, concierge services, spas, banquet facilities, meeting rooms, etc. Limited-service hotels are a step down, offering a smaller range of services, usually for lower room rates. The budget hotel is a very basic, no-frills, short-term facility. Extended-stay hotels, falling somewhere in between limited-service and budget hotels, offer amenities such as kitchenettes and laundry facilities, and they provide discounted rates to encourage longer stays of anywhere from a week to several months.

Another subset of the hospitality asset type is flagged or unflagged hotel chains. Flagged hotels are properties carrying branded names, such as Comfort Inn, Hilton, Holiday Inn Express, etc. Unflagged hotels are owned and operated by independent parties and not affiliated with any larger brand.

Other real estate types

There is actually a sixth type of real estate, one that is not likely to appear very often in the crowdfunding space, at least not in the near future. These types of property, which really do not fit the mold of the five major real estate types, are more cultural in nature. They are places like sports venues or facilities, such as golf courses, ice rinks, and marinas. The sixth type also includes religious facilities, such as churches, mosques, and synagogues.

This sixth segment of real estate is highly specialized. It often has unusual characteristics and distinct financial structures that, incidentally, Chapter 8 on the eight financial keys will also unlock for you.[6]

Class A, B, and C real estate

For the most part, each asset type we just covered can be segmented into Class A, B, or C. Those classes are driven by different standards of quality, such as location, available amenities, age of the property, condition of the property, and caliber of the tenants—though these can be subject to interpretation.

To be clear, as with anything in real estate, nothing is understood uniformly by everyone across the industry. There are a lot of gray areas within real estate, places where definitions and descriptors are unclear. You should certainly do your own investigations and drill down to understand whatever is being told to you. You want to make sure it is consistent with your own understanding. It is very important that if you have any doubt at all about what you are being told about building class, you check how this is being defined. If it is not clear to you, do not be afraid to ask.

That advice holds for any term used in this book or that you might come across in real estate. There are very few standard definitions, just as there are very few terms in art or industry that are uniformly understood. You need to understand the basic concepts and be sure that the developers with whom you are considering making an investment share your understanding.

A wide range of differentiators are used to determine building classification, and these are covered in the following sections.

Class A

Class A buildings are the highest-caliber buildings within an asset type. These are found in the best locations. Typically of high-end construction and with high-end interior finishes, Class A buildings often have modern or elegant architectural design. Also, they have a variety of what might be considered luxury amenities, including valet services, recreational facilities, and top-rated food and beverage offerings. In short, Class A buildings are best of class in their area.

While historical buildings can be Class A, often they are not, primarily because of age and because the systems inside them—including the fenestration (windows), heating and ventilation systems, ceiling height, and other features—typically do not meet modern standards. So it can be difficult for historic buildings, as beautiful as they are, to fit into the Class A category, though this is not entirely impossible. In many cases, they fall into a category of their own.[7]

Class B

Class B products are typically older buildings, though it is possible to build a new Class B product. Tenant mix influences the class status of a particular building. A Class A building might have multiple floors leased to a very high-credit institutional tenant. A Class B building, on the other hand, may be subdivided into dozens of smaller units with non-credit tenants.

Class B buildings sometimes have the potential to become Class A, and some sponsors specialize in exactly that.[8] Oftentimes, however, simply because of structural restrictions—maybe because they are old and just not configured correctly or because they have irredeemably low ceiling heights—they will be forever Class B or, if not taken care of, could even become Class C.

Class C

Class C is really at the bottom end of the scale within each of the five types of real estate. These are barely functional and, indeed, may be functionally obsolete. For example, a warehouse with 18-foot ceilings may have been considered state of the art when it was built, but is significantly below today's minimum standard for warehousing.

Class C buildings have low rents and a lot of deferred maintenance. The building might be nearing the end of its useful life, which makes it a prime candidate for demolition and complete replacement, or for adaptive reuse—like taking an old warehouse in a downtown location and converting it into loft apartments.

Impact of tech

One of my private clients develops office space for some of the tech companies in the San Francisco area and in Silicon Valley. He explains that what he is building today is expected to be functionally obsolete within five years. That is because the technology built into these properties rapidly becomes outdated. The way many buildings are being constructed now involves creating a shell with a long life span where the building's interior—the tenant improvements—is expected to be ripped out and completely rebuilt within a relatively short period, simply because technology will require a different design.

I discuss this in my podcast with Bob O'Brien, Vice Chairman at the accounting firm Deloitte. He talks about how technology is changing the way buildings are being viewed and developed. He discusses at length how, for example, parking may become an obsolescent feature of a property because of the advent of autonomous vehicles (as mentioned above).[9] The takeaway is that as time progresses, buildings can move from one class to another because of their configuration and construction. Today's Class A building could be tomorrow's Class B or even Class C building.

Investment-grade real estate

Class A is often considered to be investment grade, which means these buildings attract the wealthiest investors, like the biggest pension funds, the endowments, the ones that want to take on the least amount of risk. They want to coupon-clip[10] with a predictable, consistent stream of income. Sometimes Class A properties are considered to be 'trophy properties.' Trophy properties are those that hold some additional value beyond the purely economic. They are likely Class A properties, though not always, but are generally exceptionally well located, architecturally noteworthy, possibly of historical significance, and perhaps have spectacular views or extraordinary amenities. They command higher prices than other properties because of the prestige associated with either being a tenant or ownership.

In the middle strata, after the investment-grade trophy properties, sits institutional properties. These are properties of substantial scale. They also attract the largest investors, but they may contain some value add component, meaning that improving certain aspects of their current condition to upgrade quality can generate higher returns for investors. For the most part, they contain a stable tenant base, and often they are located in major metropolitan areas, but not necessarily only in primary cities.

At the bottom of the scale, Class C is for speculative investment. These buildings emphasize functionality. Most of these properties are outdated and provide a less favorable tenant mix, and they often have high vacancies. They are not necessarily large buildings, and they are of such inferior quality that they seldom attract the attention of large investors or institutional endowments or pension funds, etc. This is the mom-and-pop class of commercial real estate.

An expanded look at asset classes

Let's drill down a bit more on the three different real estate classes. The Building Owners and Managers Association (BOMA) International has some great resources.[11] BOMA International has created two different classifications. The first is metropolitan-based definition—Class A, B, and C—and the second is international-based definition. These distinguish between investment, institutional, and speculative buildings.

As you start to search for deals to invest in, you will find that most crowdfunding real estate sites offer either assets in transition or, in some cases (like with Feldman Equities), already near stabilized Class A product; that is, a core plus strategy. (We will talk about core plus in more depth in Chapter 5.) That is, they offer investment opportunities in what will become stabilized office properties. There are very few crowdfunded real estate sites that offer Class C options. If they do, it is only when a robust business plan is in place that indicates a strategy to move the building into Class B or better.

Certainly, in the vast majority of crowdfunding real estate deals at the moment, what you are looking at is some kind of transitional quality, where you take an asset from one class and step it up to a better class with the goal of increasing revenues and value.

Correlation between asset class and returns

Curiously, when it comes to investing, there is actually an inverse correlation between asset class and returns. The higher up you enter the real estate investment game, in terms of the class of the investment asset, the lower the risk; consequently, the returns are likely to be lower. The lower down the asset class scale you enter with your investment, the higher the risk that you might not be able to elevate that building from Class C to Class B, or Class B to Class A. The risk is higher, but the returns are commensurately higher as compensation for that risk.

A friend of mine in Los Angeles who owns upwards of 10,000 apartment units—probably worth $1.4 billion or so—likes to say, "If you want to live with the classes, you have to work with the masses." Whenever he invests, he likes to go into really tough areas and buy in to Class C projects, upgrading them to Class B or even Class A for the location. In so doing, he takes a lot more risk, but he also captures a much higher return than just buying assets that are already Class A.

Conclusion

That covers the five types of real estate and their respective class distinctions. We have discussed office, industrial, retail, residential, and hospitality. It is a good idea to narrow your focus to one of the five types and begin to specialize. When you get to it, the chapter on the eight financial keys to real estate investment will unlock the doors to understanding the financial aspects to any of these. The keys are consistent across all asset classes and all types of real estate, and they allow you to compare one against the other.

Since we have now covered the five types of real estate, this is the perfect opportunity to visit some of the crowdfunding websites. Have a look at the deals they offer. Are they offering every one of the five real estate types, and what is the range of classes you are finding across the different types?

As I've mentioned, you should take the opportunity now to hone in on those real estate types and perhaps the classes of real estate that you feel most comfortable with. Go see what you can find. There is no time like the present. Try to identify some deals, and as we progress through this book, you will be able to drill down further on the deals you like and decide which ones are worthy of your investment.

Notes

1 Chris Loeffler, CEO at Caliber, talks about crowdfunding in my conversation with him, here: https://gowercrowd.com/podcast/216-chris-loeffler-founder-ceo-caliber-companies
2 Bob O'Brien, Vice Chairman at Deloitte, talks about parking trends in my conversation with him, here: https://gowercrowd.com/podcast/robert-t-obrien-vice-chairman-deloitte
3 You can read about Shopping for the Truth here: https://partners.wsj.com/icsc/shopping-for-the-truth/maximizing-the-human-experience/ Or you can visit the International Council

of Shopping Centers website to learn more about this and other relevant retail industry topics: http://ICSC.com

4 You can listen in to the fascinating conversation with Professor Syverson here: https://gowercrowd.com/podcast/201-will-the-crowd-fund-the-next-generation-of-retail

5 In our conversation, Chuck describes his philosophy to real estate crowdfunding and why a huge institutional company like KBS would choose this avenue to raise money. The conversation is available here: https://gowercrowd.com/podcast/222-chuck-schreiber-ceo-co-founder-kbs

6 While I am sure there are many other types of real estate, a potentially major one just recently floated across my desk—agriculture. Everyone says apartments are such a good an investment option because "we all need to live somewhere," but Chris Rawley of Harvest Returns, who I spoke with for my podcast and who runs an agriculture syndication platform, echoes that sentiment when he says, "we all need to eat something." Check out our talk here: https://gowercrowd.com/podcast/chris-rawley-founder-ceo-harvest-returns

7 I am a big fan of historic buildings and had the pleasure of talking to Professor Michael Tomlan, at Cornell University, about the origins of the historic preservation designation in American real estate. You can listen here: https://gowercrowd.com/podcast/009-historic-preservation-the-penn-central-case

8 Feldman Equities, out of Tampa, Florida, are just this sort of value add office sponsor. They specialize in finding substandard skyrise office buildings that can be redeveloped into institutional-caliber Class A buildings. Find out more about how they do that in my conversation with the CEO, Larry Feldman: https://gowercrowd.com/podcast/larry-feldman-ceo-feldman-equities

9 See Note 2.

10 The term 'coupon-clip' is a holdover from the days before crowdfunding was prohibited by the 1933 Securities Act. Shares in an offering that carried with them interest payments would be issued together with redeemable coupons—in paper format. A shareholder would submit the coupon which retained its value based on the original price of the shares to the issuing bank to redeem their interest payment. As the price of the shares fluctuated, the value of the coupon in percentage terms would vary relative to what a shareholder had paid for the underlying shares. The idea of coupon-clip real estate investment is that there is nothing to do but to collect interest on the invested capital.

11 Find out more at BOMA International's website: www.boma.org/

5 Four development strategies

This chapter introduces the different strategies that developers employ to invest in commercial real estate. Understanding these different investment options will help you explore ways to diversify your portfolio to mitigate risk and maximize returns.

The four development strategies are core, core plus, value add, and ground up, and the main differentiator between them is their risk-return profile.

Risk and return

The fundamental rule that applies to development strategies is the risk and return rule: The higher the risk, the higher the returns; the lower the risk, the lower the returns.

During my second podcast series, I spoke to many of the founders and CEOs of the major real estate crowdfunding sites. For those sites that held webinars, they told me that there was a positive correlation between the number of people who registered for and showed up at a webinar and the advertised IRR for the deal being pitched in the webinar.

This is a paradox that has emerged as a result of real estate crowdfunding and the visibility that deals have online. If IRR is the driving force behind investors' decisions to invest in individual deals, they may not be taking a critical enough view of transactional risk. At time of writing, the real estate cycle is peaking, and yet on real estate crowdfunding sites, the deals with the highest IRRs tend to win the race for investor capital. Institutional and other sophisticated investors understand that returns diminish as the cycle prolongs and set expectations accordingly; crowd investors as a group apparently do not—at least not yet.

Apart from being counterintuitive to the market cycle, this paradox also creates its own bubble. Inexperienced real estate investors and some crowdfunding platforms are signaling to sponsors that as long as projected IRRs are high relative to other sponsors, they will be successful in raising capital. Some of those crowdfunding platforms are similarly telling sponsors that without high projected IRRs, investors will not invest and that they, the platforms, will not take on projects that don't conform to investor demand. Uninformed investor demand driving sponsor actions—a case of the tail wagging the dog.

Despite knowing that lower, not higher, IRRs are more consistent with conservative underwriting as the cycle enters the hypersupply phase, sponsors have become acclimatized to raising capital from inexperienced crowdfund investors who will only invest in deals with the highest IRRs.

As a consequence, sponsors must offer higher IRRs to compete with other sponsors, which creates the temptation to be 'relaxed' on underwriting assumptions in order to ensure that IRRs are attractively high.[1]

Diversification

Besides the paradox of seeing IRRs go up instead of down as the cycle lengthens, due to the dynamics of real estate crowdfunding online, the various development strategies will have different IRRs. To mitigate risk while increasing your returns, you might want to consider a variety of development strategies with different risk-return profiles, mixing some of the low-return/low-risk options with relatively high-risk/high-return strategies to avoid exposing your entire portfolio with just one sort of profile.

Historically, a developer would ask investors for a large minimum investment. This sum could be a few hundred thousand dollars up to millions of dollars in a single deal. Investing such a large sum of money in a single deal exposes the investor to high risk concentration. With the advent of crowdfunding, minimums have dropped dramatically, allowing for considerably more portfolio diversification than has previously been possible.

This flexibility makes crowdfunding an appealing option for both new and seasoned investors, because crowdfunding real estate investments ensures that you don't have to put all of your financial eggs in one basket.

The big four development strategies

As I've mentioned above, you can break real estate down into some fairly distinct components. There are five types of real estate, eight phases of development, eight financial keys, and—the subject of this chapter—four development strategies.

Remember, specializing in one strategy is a good tactic for real estate investment. To get started, choose one real estate type that interests you the most and focus your energy there. Consider concentrating on one particular asset class and drill down on one of the development strategies that we'll discuss below. Focusing on one area will make it easier to keep an eye on your investments and streamline your energy for the best results.

You'll find developers who focus primarily on apartments and others who focus on office buildings. Typically, these developers don't cross over into other real estate types. Office developers may be aghast at the idea of owning apartments, because they can't stomach the idea of having somebody living in a property they own. Equally, many apartment owners are flummoxed by the idea of having to rent to a business rather than to an individual. Apartment investors are much more comfortable doing credit checks, conducting background checks, and taking deposits from individuals than they are trying to figure out whether a business is a creditworthy tenant. Similarly, most developers specialize in one of the four primary development strategies.

Figure 5.1 shows a classic risk-return profile. On the bottom left, with lower risk/ lower return, you have the core product—the first development strategy we will look at. Next on the risk-return line is core plus, a slightly higher risk/higher return option.

Moving up the risk/return scale are value add and ground up development. As I've mentioned, there isn't a central standards authority that every developer must adhere to— I've seen the terms on this graph referred to in multiple ways depending on the source. You may, for example, hear value add or even ground up described as being opportunistic, depending on the background to a deal, especially if there was some form of 'distress' associated with the original sale of the asset.

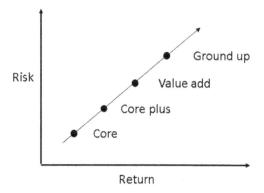

Figure 5.1 Risk versus return

Core

Core investing is sometimes referred to as a 'coupon-clip' strategy. The idea is that someone investing in core real estate is trying to minimize risk as far as possible by simply looking to make a yield on their investment and minimize any downside that might come from attempting to add value to a project in any way.

Core products are typically stabilized and located in prime downtown locations. They are these big skyscrapers that are also referred to as 'institutional assets.' The term 'trophy buildings' is also used synonymously with core assets; although 'trophy' has a different connotation; these are beyond core assets in that they have some historical or other type of prestige attached to them that brings additional, intangible value.

The antithesis is a run-down building on a third-rate location with a bunch of low-credit tenants struggling to pay rent. When you walk in the lobby, it smells, and there are stains on the floor. That is not a core asset that an institutional investor would have the slightest interest in. These lower-grade assets are not generally the kind of asset that a big insurance company or a pension fund or a major university endowment would invest in.

Consider for a moment the motivation of an institutional investor, like a pension fund, when they look for buildings and why core assets would be their preference. A pension fund is in the business of collecting contributions from their members and investing those funds over many decades, because they expect eventually to have to pay retirement pensions to their members. Members may come in to the fund when they are 25 years old, and whatever those folk pay into the pension fund needs to be paid out 30–40 years later. Life insurance companies work in a similar way, collecting premiums and only paying out after a multi-decade cycle.

Because of this, big institutions typically invest in core assets, as these are viewed as offering very stable cash flow over a long term. Even if the yield is not tremendously high—indeed yields on core assets are generally the lowest of all real estate asset types—you know that it is going to be steady and stable. Life insurance companies, pension funds, and endowments don't generally like to take risks. What they value the most is stability and predictability.

As an investor in a crowdfunded deal, when you invest in a core product strategy, you are investing in a completed building with great tenants, and the biggest responsibility the sponsor has is to collect the rent and maintain and manage the building. While these are by no means easy tasks to perform well, there really is nothing else to do.

Core products are usually located in major downtown metropolitan areas, like New York, San Francisco, Los Angeles, Chicago, etc. These heavily populated areas have diverse employment bases and are supported by easy access to residential areas—increasingly located in the immediate vicinity.

Properties usually have high, stable occupancy levels with low vacancy rates and host high-credit tenants—household names or leaders of the industries in which they are active. These tenants may have been assessed by credit ratings analysts, and this enables the owner to take that credit to investors and to banks.

Core product rents are typically at or very close to full-market rents. With no lease-up to perform (i.e. as the buildings are at full occupancy, there is no need to engage in leasing activities), all the investor has to do is manage any rare instances of turnover. Since these buildings have long leases with staggered expirations, a core product offers tremendous stability. In exchange for that stability, however, there's limited potential from a capital gains perspective, since there's little opportunity to add value to the product.

Core products are often the last to suffer when there's an economic downturn, because tenants are the best-of-class in their own industries. While the tenants may be impacted by an economic downturn, they're less likely than other businesses to go bankrupt, downsize, or otherwise default on the terms of their leases and move out. When you purchase a core product, you're buying it primarily for the yield that you will earn—in many ways like a treasury bond but without the government guarantee.

Interestingly, core products, because of their sheer scale, have up until very recently been out of reach for all but the wealthiest and largest investors, such as high-net-worth families, institutional investors, pension funds, endowments, sovereign wealth funds, etc. Real estate crowdfunding has leveled the playing field somewhat by bringing to market investment opportunities that have never before been available to any but the largest institutions.

Today, with the emergence of real estate crowdfunding, investors have access to a variety of funds, where they can invest as little as $5,000 to $10,000 for a piece of the action. And there are many opportunities to invest in these core assets, in exactly the same manner as the wealthiest and largest investors have been doing since the 19th century.

Core plus

The idea of core plus investment is to identify a building that has the potential to be a core building and upgrade it so that it achieves a standard sufficient to attract institutional investors.

Put another way, a core plus product is a core product in need of some redevelopment. With a core product, investors collect steady income on a very stabilized, low-risk product. With core plus, the sponsor is looking to make some type of improvement in the building to increase the value. When you invest capital in a core plus product, you're not just looking for yield; you are also seeking capital appreciation.

Sponsors investing in core plus properties are entering the development process earlier than those who invest only in core buildings. They will buy big buildings in a Class A location that are somewhat run-down physically and that are underperforming in that their lease-up is distressed. A core plus investor is not afraid to buy a building that is 30 percent leased provided it has the potential to eventually become a Class A core asset, because their ultimate goal will be to attract an institutional investor to either provide permanent finance, buy the building, or partner with them.

There are a few factors that can cause these properties to fall outside the core investment category. Typically, something is amiss with the properties. They might have high vacancies. The tenants may not have the highest credit ratings, which could indicate that they are high-risk tenants. Additionally, lease terms might be shorter than those for core properties, so there is potential for a higher turnover of tenants. This makes it difficult to get the most favorable loan rates and terms from banks, who look at the weighted average lease term (abbreviated to WALT) as part of their underwriting of the creditworthiness of a building.

There are plenty of ways to improve the value of these properties, like improving the common areas, changing up the tenant mix, increasing rents, extending the length of existing leases, and filling any vacant spaces.

With core plus products, adding value means spending capital—in addition to acquisition costs—to make necessary improvements. Cleaning up the lobby, renovating washrooms, landscaping outdoor areas, negotiating new leases, and paying commissions are some of the things that require capital expenditures, but could increase the building's value.

When sponsors purchase core plus products, they're paying prices similar to those of core products; but the expectation is that through adding value to the building, those yields will increase significantly.

An example of this is the Wells Fargo Center, a 390,000 square foot office skyrise in downtown Tampa, Florida. The building was acquired by Feldman Equities in partnership with Goldman Sachs in 2012 for $44 million. At the time of acquisition, the building was about 70 percent leased. Although the infrastructure of the building was okay, there had been no reinvestment in the building since 1987. It was underperforming financially because physically, in terms of how the building appeared to tenants and visitors, it looked like it had been frozen in time. Multiple reasons were given for the building not being leased, but the bottom line was that nobody had looked at the building and planned a complete renovation until the Feldman/Goldman partnership in 2012.

The new owners came in and completely renovated the building from top to bottom. They installed a brand-new high-speed elevator system and created a 24/7 'chill zone' with leather couches, special chairs where users could attach their personal communication devices, and a cappuccino machine. They built a fitness center of a standard you would only expect to see in an expensive membership gym, which tenants could access almost for free. The new owners built out a brand-new common-use conference room for smaller tenants to access when they needed it, thus avoiding the need for separate conference rooms in tenants' own spaces that they would have to pay rent for. In the lobby, they included seating areas for tenants and visitors to use their laptops, installed high-tech lighting, and included 12-foot flat-screen TVs showing 'digital art.' The owners added other kinds of amenities into the renovation, including a shuttle system to transport tenants around town.

After spending some $5 million on the building upgrades, over a two-year period, occupancy went from 70 percent to 93 percent, and NOI went from $3.5 million to around $6 million. The 'risk' capital investor, Goldman Sachs, exited the deal, and at time of writing Feldman's financial partner in the building, New York Life Insurance, has a completely different investment profile. They were not trying to hit a home run on the building, but saw it as having some modest upside and as a fairly safe long-term investment. It is unlikely that an insurance company would have come into the deal before the renovation.

A core plus building, then, is one that an institutional investor would not otherwise buy, but which could provide an exit for investors once improvements have been made.

Value add and ground up

Higher on the risk-return scale, value add projects look to deliver a high return through taking on higher risk. Value add products often involve heavily flawed buildings with plenty of room for improvement. Making those improvements captures the value inherent in the building's previously unrealized potential. A value add product seeks to earn a profit through significant work on a property to reposition it, substantially increasing rents and occupancy rates.

The final of the four development strategies, ground up (sometimes called 'opportunistic'), reflects the highest risk/return. This type of deal includes assets with serious flaws that can gain significant value through improvement or redevelopment into a completely new use.

These 'adaptive reuse' types of investment strategy take the longest amount of time and have multiple layers of work to complete before investors can earn a profit, yet they can create esthetically beautiful results that dramatically enhance a building's functionality and value. For example, a former office or industrial building that has fallen into obsolescence could be converted into residential units, like high-end lofts. The property may have features like high ceilings, exquisite windows, concrete floors, and exposed beams and pipes that can make for interesting design attributes in a residential environment.

Investing in these kinds of distressed properties does pose a much higher risk and requires substantially more work (including working with city officials for proper zoning) to increase value, and then reap the rewards.

Sponsors may find other value add opportunities while they are inspecting properties and sites they considering acquiring. Look for ways sponsors are adding value by, among other things:

- adding square footage to the property;
- reconfiguring the building to increase unit numbers out of existing square footage;
- constructing a parking structure adjacent to an office building sitting vacant primarily because parking is an issue in the neighborhood;
- controlling operating costs;
- creating open working space by redesigning the interior of a building;
- innovative marketing; and
- upgrading technology.

The expectation for a value add or ground up investment is a significant increase in property value, and overall returns, once the business plan is implemented successfully. Yet, while there is a lot of potential for a major increase in returns, there is also greater risk. Value add deals require detailed business plans that outline the proposed work, and there are more opportunities for the deal to deteriorate through taking on construction, market, and timing risks.

Nearly any type of real estate can have value add potential. Even downtown skyscrapers fall into disrepair or become outdated, offering opportunity to substantially improve or upgrade them. There are many reasons a property would fall into the value add category:

- It may involve buying a completed building in dilapidated condition.
- The property could be partially complete. When a developer runs into financial problems, they may be unable to complete a project, leaving a partially finished building.
- Value add properties often have high vacancy rates.
- The building could be suffering from poor management.
- There may be considerable deferred maintenance.
- Tenants' credit could be poor.
- Lease lengths may not be optimized.

Some value add opportunities may appear counterintuitive, and these require the seasoned eye of an experienced sponsor to identify them. For example, at first glance you may think that an apartment building with 100 percent occupancy and a waiting list for units is a good sign. But this might be due to poor management and rents that are below market.

I looked at acquiring a 104-unit condo development project just like this. The economy had taken a downturn, and the developer had pivoted from a for-sale condo development project to a for-rent apartment project in order to save the deal. The problem was that they had no experience in apartment rentals. They managed to get the property to 100 percent occupancy and kept it there for a number of years while managing an extensive wait list for units. In their mind, they were doing very well, but we could see that rents had significant potential to rise.

Another way I've seen investors add value to apartment buildings is using the inspection, valuation, and due diligence phases to find ways to reduce operational costs—savings that can drop straight to the bottom line, adding directly to NOI. For example, this might include reducing utility expenses such as water and power consumption by upgrading the mechanics of a building or by passing the utility expenses on to tenants in the form of a ratio utility billing system.[2] While simple, these types of savings add straightaway to NOI, and this adds value to the investment because any operational savings are, effectively, new revenue streams.

As already mentioned, sponsors may also create additional revenue streams by adding more facilities or amenities that are attractive to tenants. One amenity that may attract tenants willing to pay a little more is additional parking, or premium parking structures that offer easy access to buildings for an extra fee. Adding laundry facilities to an apartment building provides a convenience for tenants and adds another revenue stream. Trion Properties, a value add apartment developer in Los Angeles that invests in properties up and down the west coast, will typically install vinyl plank flooring, stainless steel appliances, new fixtures, and cabinets, remodel the bathrooms, and then paint the exterior a bright color to give it some curb appeal. To attract further attention, Trion will add value by creating property-specific websites, where residents can pay rent and submit maintenance requests online from their cell phones or computers.[3]

When I was president of the Asia-Pacific region for Universal Studios during the 1990s, we used to joke about all the various ways we could bring additional revenue to a building. The properties I was developing cost in the tens of millions of dollars, and the president of the studio at the time liked the idea of adding small water features into the lobbies of these venues. He figured that because people threw pennies into these water features, it would add to revenue. He was just kidding of course, but finding creative ways to add value is a seasoned sponsor's primary function.

Other things can significantly affect an asset's value. For example, a shopping center that loses a major anchor client could be left with a giant retail space that's difficult to fill. This type of vacancy creates a value add opportunity. An investor could reconfigure the empty anchor space into smaller units for sublease to smaller tenants, which could allow the investor to earn even more than the original tenant was paying.

Overleveraged properties offer another value add opportunity. During the economic downturn that started in 2007, I saw a lot of instances where sponsors had borrowed too much money and then lost tenants who could no longer afford their rent due to the shifting economy. The loss of these tenants made it impossible to keep up with debt payments. Some sponsors stopped making payments on their loans and/or stopped spending money to improve, or even to maintain, their properties. This led to properties becoming increasingly neglected while maintenance issues accumulated, tenant quality declined, and vacancies increased.

When overleveraged properties fall into disrepair and enter foreclosure, it creates a wealth of value add opportunities for investors to buy in to. Investors can spend money performing all of the deferred maintenance and repairs left behind by the prior owner and improve the value of the property so they can earn higher returns.

Sponsors can also add value to the investment by restructuring the capital stack, refinancing, or recapitalizing a building in order to reduce the debt burden or to return equity to investors.

The key takeaway from this chapter is that the more work that needs to be done to a property, the higher the risk that the business plan will not be executed as anticipated. To compensate for this, returns should be commensurately higher, and it is this concept that is often referred to as the 'risk-return profile' of a project and is what sponsors are referring to when they talk of risk-adjusted returns. Bottom line: If you see a project on a real estate crowdfunding site or platform that has high projected returns, always assume that the risk of loss is equally as high.

Notes

1 I have written in detail about this paradox and other impacts of crowdfunding on real estate in an article on my website, 'The impact of crowdfunding on real estate waterfall structures.' You can access that here: https://gowercrowd.com/learn/impact-of-crowdfunding-on-real-estate-waterfall-structures

2 A ratio utility billing system (known also as RUBS) is a way of calculating a resident's utility bill based on occupancy rate, number of tenants in a unit, apartment square footage, or other metrics, and then passing on the proportional share of utility use to the tenant. It is used when the installation of submeters is prohibitively expensive.

3 You can hear Trion's Managing Partner, Max Sharkansky, discuss his approach in my conversation with him, available here: https://gowercrowd.com/podcast/315-max-sharkansky-managing-partner-trion-properties

6 The eight phases of real estate development

This chapter covers the eight phases of real estate development. The good news is that as a passive investor you need only focus on the financial concepts that underpin this eight-phase development cycle, and in this chapter we'll go through all eight phases so you can contextualize the financial terms you need to know.

The eight phases of development

There is a critical path in real estate development that gets every deal from its natural start to its natural conclusion. So to successfully execute on a real estate business plan, there is a sequence of events that occur in a strict chronology. If the order in which these events occur do not follow the critical path, the cost of developing the project and the time it will take to complete will increase.

To help you understand how the critical path works, let's take a look at the eight phases of developing a property, all of which must occur in the following order.

1. Find a property to buy

The first phase is finding the asset, the property that you want to develop. There are many ways a sponsor might find an asset to invest in. They can look on and off market for deals through their broker networks. They can tap into their own private lead source networks. Particularly proactive developers will conduct research and cold-call prospective sellers, acting as their own broker in the transaction.

2. Formulate the business plan

Once a sponsor has located the property they want to develop or buy and improve, they then enter the second phase. This involves deciding what exactly will be done with the property and then creating a strategy for implementing the plan. This phase is the road map, the business plan, for transforming the current property into the property they want it to become.

3. Conduct a feasibility study

The next stage is determining the feasibility of transforming the desired property, according to the business plan. This requires drilling down on the underlying assumptions to ensure that the business plans isn't pie in the sky. Typically, while the feasibility phase does require precision, this is simply an initial screening to ensure that further effort is actually warranted. Sponsors use basic indicators to determine whether a project is feasible. You can think of this as the 'back of an envelope' phase of real estate underwriting. Key metrics, like cost per square foot, assumptions regarding projected rents, finance terms, etc., will be used to gut check the assumptions going in.

4. Enter contract negotiations

If a project looks to be feasible, the sponsor will then enter contract negotiations with the owner of the property or the owner's agent or broker. The goal of this phase is ensuring that there's enough time to verify the assumptions that were loosely underwritten during the feasibility stage, the initial look at the business plan, and to reconcile those assumptions with the financial assumptions as negotiated in the contract with the seller. In this phase, the sponsor verifies that the negotiated price, terms, and conditions are consistent with being able to execute on the business plan.

5. Conduct thorough due diligence

After contract negotiations and once the buyer and seller have agreed to all the terms and conditions of the sale, it's time for the heavy lifting: due diligence. During the due diligence phase, the sponsor drills down on every single assumption in the business plan to ensure they have covered all the bases. By reviewing everything with a fine-tooth comb, they won't be tripped up by unseen landmines, that could add to development time or cost, when they execute on the business plan. The reason why it is so critical to understand the financial keys you're going to learn about later in this book is that they drive the due diligence process. There are two key components in due diligence. The sponsor must:

1. build a financial analysis of the project in a spreadsheet;
2. justify every cell of their spreadsheet.

Each of the numbers the sponsor plugs into the financial analysis must be backed up by rigorous research and analysis.

6. Make a formal commitment to buy

Once due diligence is completed, and the assumptions have been validated and ratified, the sponsor is now in a position to formally commit to the project. Up until this point, the sponsor can walk away from the deal without risking any capital. Even though they put down earnest money (the commercial real estate equivalent of the deposit in residential real estate) as part of the contract negotiations to demonstrate goodwill, intent, and ability to close, under most circumstances those moneys are fully refundable before a formal commitment to buy is finalized.

As a result, the formal commitment stage is the point of no return for earnest money. This is when the sponsor informs the seller that the deal is going forward and the earnest money becomes nonrefundable, the purchase and sale period ends and the sponsor brings all funds necessary to close, and the property ownership transfers to the sponsor.

7. Commence development

Once the sponsor has underwritten all the assumptions, owns the property, and is prepared to start executing on the business plan, the development phase begins. Unless the business plan calls for holding the property indefinitely in the eighth phase, development typically takes more time to complete than any other phase.

8. Sell it or manage it

In phase eight, the sponsor has completed the business plan and is now left with only two choices—either sell the property or manage it and collect rents. This choice is driven by the business plan and demands of investors relative to their rights according to the agreement they have with the sponsor. Even if a business plan called for one form of exit once completed—a sell or hold scenario—circumstances may call for modifications and adaptation, depending on market conditions and other factors.

Bifurcating the development process

The eight phases of the development process can be split into two key stages. The first stage occurs before the sponsor commits to the project, and the second stage occurs after the commitment to acquire the property has been made. These stages are broken as follows.

- Before commitment
  1. Locate an asset.
  2. Design a business plan.
  3. Conduct a feasibility study.
  4. Enter into contract and term negotiations.
  5. Execute a thorough due diligence process to ensure the accuracy of all assumptions.
  6. Line up finance and prepare to close.
- After commitment
  7. Execute on the business plan.
  8. Hold or sell the finished building.

In most cases, prior to being invited to participate in a deal as an investor, the developer will have already concluded six of the eight phases of development. Once you have made your investment, typically there will only remain the task of executing on the business plan, culminating with the completion of the project by sale or ongoing ownership and management of the property.

Sponsors are specialists

It is generally better for you to focus your attention on sponsors who have specialized in one particular asset type. It is very rare to find a generalist in real estate development, someone who has mastered a variety of different asset types.

I occasionally come across some exceptions, as I've mentioned already in this book. These kinds of sponsors are generally inexperienced but have been bitten by the real estate bug. They sound like a kid in a candy store. They know that they want to be in real estate, but being unable to decide what asset type to focus on, they chase any deal that seems to make sense. This is a very hazardous environment for an investor.

It is extremely difficult to understand the intricacies of any single real estate type, let alone become a master of multiple asset types at the same time. Your best bet is to find subject matter experts, seasoned real estate professionals who can demonstrate through their background and experience that they are masters in a specific asset type. You can use the knowledge that you are gaining by reading this book to identify which of the sponsors out there are true experts, and then you can invest in a range of different asset types by partnering with only the best of the best in each category.

The 'before commitment' process

Finding a deal is the first step a sponsor must engage in to be in business (duh!). Once found, the sponsor can then compare the deal with other available options and draw upon their experience and knowledge to determine if it is worth looking at more closely. They will examine multiple deals simultaneously, comparing options and conducting desktop analyses. They look for deal killers—factors that eliminate a deal from consideration—by performing feasibility studies, repeating this process continually while they review a large pipeline of deal options until they locate a favorable deal that meets all the criteria for a successful outcome.

Once a sponsor locates a feasible and potentially successful deal, they then formulate an offer which they present to the seller, after which negotiations with the seller begin. During this time, both parties ensure that pursuing the deal is worthwhile. If negotiations are successful, the sponsor will then contract to buy the asset.

Typically, this gives the sponsor an exclusive right to buy the property for a certain period of time, during which the due diligence phase begins. In this phase, the sponsor conducts thorough due diligence, drilling down on all the assumptions that led them to believe the property was worthy of acquisition and development in the first place.

The deal-finding process resembles a funnel, with the number of deals coming in from numerous sources being much larger than those deals that actually get accepted. At the bottom of the funnel sit the deals that make it all the way through the due diligence process—the point at which the sponsor starts to search for potential deal killers and to put substance on the business plan by lining up a team to execute if the deal goes ahead.

During financial analysis and property research, the sponsor conducts rigorous underwriting of their assumptions about these deals, searching for all possible factors that may kill a deal. Earnest money funds are not at risk during the due diligence phase. In general, there are numerous factors that kill deals, almost always pertaining to disproving the original financial thesis and underlying assumptions. Should any of these factors make the deal unacceptable, then the sponsor will either attempts to renegotiate the deal or rejects it, contracts are exited, and the seller refunds any earnest money to the sponsor. The entire process is repeated by the sponsor as they search for deals, and it keeps on repeating until a viable deal is found.

Once the due diligence and financial analysis of a deal pan out and all underwriting assumptions are validated, the deal gets the green light. At this point, the parties execute a purchase agreement and make a commitment to move forward with the deal. The sponsor fully funds the transaction and takes ownership of the property at closing. They are, at this point, fully committed and ready to carry out the business plan.

The 'after commitment' stage

In the second stage of development process—after commitment—the sponsor commences execution of the business plan. They will have already arranged financing and debt will be in place, using your and others' equity investment to close on the deal or to close with their own capital and then backfill with yours, and drawing on debt as approved expenditures are incurred.

A very important question to consider as you examine deals is: Has the sponsor already bought the property, or is it yet to be acquired? If it is the latter and you are being asked to invest before the closing, what assurances are there that the deal will close?

After taking ownership of the property, the sponsor obtains all the required entitlements and permits. Just about every municipality within the United States has its own local planning and building and safety permitting requirements, that the sponsor must adhere to. This is not a trivial phase of the development process, and another question that you want ask is: Does the sponsor have the necessary permissions to do what they are planning to do, or must they still endure a discretionary approval process that could involve both the municipal authorities and the local neighborhood groups?

Presuming that a project is approved for construction or redevelopment, the sponsor's job is to hire and manage contractors to perform the work, and a project management process is created to manage construction or improvements. This is to ensure that they perform all jobs as specified and meet all budget and time requirements.

After completing the property improvements, the building may be leased up or sold. In some cases, tenants are recruited to take space during the construction phase; or they may lease space post completion (obviously, no one moves in until the building is completed). The leasing process continues until stabilization occurs, which is when the property is filled with long-term rent-paying tenants. At this point, the sponsor either elects to sell the property or hold it for income. Again, this is something that occurs in the 'after commitment' stage of development.

To reiterate an earlier point, as an investor, you're not developing the project yourself; the sponsor handles all aspects of the development function. As an investor in one of these deals, you should understand this important distinction because it leaves you with only one vital thing to do, and that is to double-check the sponsor's financial assumptions.

Finance drives real estate

The context for every single real estate deal is finance. Without it, nothing gets built. Real estate needs capital, which is the reason you are involved: the real estate sponsor wants your money.

To understand a real estate deal, you must drill down on the money. Follow the money in precise terms, and you will be able to understand any deal. In that context, let's examine the background to real estate deals in terms of who gets paid, what they get paid, and why they get paid.

We'll do this by looking at the three primary parties connected to the cash—the bank, the investor (you), and the sponsor. Below we examine each key party's financial interest in seeing the deal come to successful fruition.

The bank

Out of the three primary parties, the bank is the party that takes the least risk. This may seem counterintuitive, because it's the bank that is usually on the hook for the largest dollar amount. Although they typically provide a significant proportion of the money— 60 percent, 70 percent, 80 percent—for a deal, they take the least risk because they retain the greatest control over the property if things go wrong.[1]

Banks are paid for their loans through interest income based on the loan amount, in addition to loan fees for putting the loan package together and other ancillary servicing fees. Typically, banks are also reimbursed for any out-of-pocket expenses, such as legal fees, document fees, and appraisal fees. Most importantly though, the bulk of the bank's earnings come from the interest rate earned for extending credit to the borrower in the form of a loan.

While it certainly isn't definitive of the caliber of a deal, one factor worth examining is the type of lender participating in a deal. Knowing who the lender is may help shed light on the quality of a deal. For example, if it is a major financial institution that is lending on the project, this may indicate a higher lending standard that requires a more diligent underwriting process. On the other hand, non-banks or secondary financial institutions might indicate a lower level of quality based on their underwriting standards.

So now we know how banks are paid; but what are they paid for? What do they do to earn that interest they're being paid? First, they lend their depositors' money. A bank only has money because depositors put money into that bank. This means the bank has a fiduciary responsibility to ensure that lending to a sponsor is done wisely and prudently. In the end, the bank receives payment for ensuring that they're making prudent lending decisions with their depositors' money.

They also earn their interest and fees for performing extensive research and under-writing to determine the quality and risk of the deal. Bank asset managers do this by examining every possible facet of a particular transaction to ensure that it meets the bank's lending standards.

The bank is also paid for compiling the legal documentation required for the deal itself. Also, it handles all of the direct and tangential processes involved in lending money to your sponsor and servicing the loan. In addition, banks are compensated for meeting compliance requirements, typically imposed by federal regulations, for how they lend money.

A bank also receives interest and fees to cover its overhead—obviously. While a sponsor may only deal with one asset manager, or one loan officer, every bank has an extensive support infrastructure behind that individual that ensures operations run smoothly.

Finally, the bank is paid to be a mostly passive participant in the actual execution of the business plan. As long as the sponsor pays them on time, the bank generally does very little in aspect of a deal. Of course, there are protections built into the loan documents signed by the sponsor that allow the bank to become hyperactive in the event of default. However, other than dispersing cash to the sponsor when they need it and processing loan payments as received, the bank mostly remains passive.

The investor (you)

The next party with a vested interest in the success of the deal is the investor (that's you). And it may come as a surprise, but you take on the bulk of the financial risk in the deal. If the deal goes south, you're the first one to lose your money. So how are you compensated to take on that risk?

In most cases, the first form of compensation you will receive will be a preferred return on your investment ahead of any other party, though always after the bank.[2] Essentially, like the bank, you receive an interest payment (the preferred return) in return for having provided capital (in the form of your investment) to the sponsor to complete the deal. The preferred return is sometimes paid on a current basis and sometimes deferred until the project is completed. In the world before crowdfunding was permitted, investor returns were typically deferred.

It is important to note that you earn a percentage of the profits, whereas banks only earn interest. To compensate you for the risk of investing in this deal, you receive interest in the form of a preferred return, plus a percentage of the profit proportionate to your investment. In other words, you get some of the back end as well.

We looked at what banks get paid for, but what are you, the investor, getting paid for exactly? What are you doing to earn the preferred return and your share of the profits from the deal, should it be successful? Well, simply stated, you're paid because you have taken on the highest risk in the deal, and you earn profits in the deal to compensate you for the risk you have taken.

You're paid because you've actively demonstrated a willingness to diversify some of your savings and put it into real estate. You earn a share of the profits because you choose to take your own safe money from the bank and put it into a real estate deal. You're also being paid a preferred return to compensate you for the risk of investing in this deal—a risk you are taking for yourself, your family, your heirs, and whatever philanthropies you're into.

All of these components justify your share of the profits, and, similar to the bank although in a different way, *you're also compensated for being mostly passive.* For the most part, once you've wired your investment capital, you do nothing. Once your money hits the sponsor's desk, you are passive and earn passive income from this deal. If it goes well, you build wealth as your investment is multiplied by some factor.

So how do you protect yourself when all you are doing is sending money to someone and then remaining almost completely passive? Well, here's a hint—it's the same way that a landlord gets rid of a bad tenant. Getting out of a bad deal starts by not getting into it in the first place. Much like getting rid of a bad tenant, the best way to do it is not to rent to them at all.

In other words, do your homework before actually committing to an investment. That is what this book is about from beginning to end: reducing the risk of investing in a bad deal and increasing the chances of getting into a good deal.

The sponsor

The sponsor is the third party with a vested interest in creating a successful deal, but what does the sponsor do to earn their share in the deal? Of course, the sponsor takes on financial risk, but they also put their time into the project. In fact, they dedicate their lives to developing real estate and building value in real estate so that they and their investors can benefit. As a result, they've made a conscious decision to take on the sponsor role.

From your perspective, the clear benefit of this arrangement is that you don't do sponsor work or development work; rather, you're paying the sponsor to do these things. Better yet, because of the way crowdfunded real estate deals operate, you can cherry-pick the sponsors who are the best of the best and invest only with them.

Taking that thought one step further, not only can you choose the sponsors who are the cream of the crop, but you can also choose a selection of sponsors. This way, you can spread your risk by creating a portfolio of investments from among the best in the industry.

Sponsors are compensated for their work in several ways. They are paid fees that cover their overhead and time, money you would otherwise spend if you were developing real estate yourself. Sponsors also earn interest on the equity they've put into the property. This is similar to how you, the investor, earns money.

The money they put into the deal is commonly referred to as their 'skin in the game.' We'll explore this in detail in Chapter 8 on the eight financial keys. The sponsor's own cash contribution to the project helps to create an alignment of interest with investors, as they are sharing in the risk of investing their own capital.

In addition to that, the sponsor receives a 'promote.' This share of the profits is paid to them once you've received your money back along with all the interest that is due to you for preferred interest. This share of the profit serves to compensate the sponsor above the fees that they earn for the work they do in developing the deal. More importantly, the promote also motivates them to be successful, because as long as the deal is structured properly, if they are successful, you will be too—and that's obviously in your best interest. In part, it is in the promote that you see alignment of interests.

Sponsors are paid to find and underwrite deals, and they are paid for dead deal costs, which cover the deals they've explored but rejected. Over time, they put at least some time and effort into feasibility studies on properties that don't pass muster, so called 'dead deals.' In some instances, as part of due diligence, sponsors spend money to conduct detailed underwriting and research that ultimately kills a deal. Their reimbursement for this is aggregated through the fees they earn in deals that move forward.

Sponsors are also paid for their time and the skills they use in negotiating a purchase. They get paid for conducting the due diligence and for underwriting every assumption in the financial analysis. Sponsors are also compensated for all development management work, for the entitlement process, and for acquiring permits required to allow construction to commence. Furthermore, sponsors are paid for the effort they put into leasing the property up and for managing the property once it is leased up. They also get paid for their presumably considerable experience as a sponsor, because such experience mitigates risk. Essentially, sponsors get paid for expertise in running a company that engages in the business of developing real estate.

Most importantly, sponsors are paid for being the most active participant in a deal. By definition, they run the project. They have the full, active responsibility of executing the business plan in the most efficient and seamless manner so that all invested participants see returns. In other words, sponsors are paid for the role they play in making sure a deal has a successful outcome.

And keep in mind one of the old clichés that still rings true in real estate—you make money in real estate not when you sell a property, but when you buy it. That's when you make the money—on the buy. And guess what? By the time you, the investor, come into the deal, that work is already done! The sponsor has already purchased the property or is on the verge of buying.

The investor's single responsibility

By the time you're considering investing in a deal, the sponsor has already completed six of the eight phases. The property is purchased. The business plan is decided. The due diligence is done. The contracts are signed. The purchase agreement is signed, and the sponsor owns the property. As the investor, you can't do any of those things. It would be futile for you to even think about doing any of this. It is not your job as an investor.

Even so, you are not absolved of responsibility. You can make sure that all six stages have been completed properly. Have the sponsors signed the contract? Do they own the property? What stage are they at in the entitlement process?

What remains are the last two phases of the development cycle—the actual development of the property and its completion and ongoing management. Other than making an investment, you won't be involved in any way, shape, or form in executing those two phases of development. So what is there for you to do?

Before you get into the deal, there really is only one thing you can do, and that is to answer the question: Are their financial assumptions accurate?

Notes

1 See the section on leverage in Chapter 8.
2 See the section on waterfalls in Chapter 8.

7 The risks of real estate investing

Before diving into the financial terms, let's consider the risks of investing, and first let's look at them from the developer's perspective. Like you, developers want to make money on their deal. But because real estate is a zero-sum game, they have an inherent conflict of interest. Any given deal can only make a finite amount of money, and that has to be portioned out somehow. Whatever money one party gets, another party doesn't get.

The sponsor also faces what is called 'the principal–agent conflict.' They are a principal on the deal, yet they are also the agent, as they have a fiduciary responsibility to act in your best interests. Herein lies the conflict. They are driven by their own best interests, but at the same time they are obligated to preserve your best interests; and in a zero-sum game, the two are mutually exclusive.

Remember, in any deal you examine, the sponsor is *selling* their deal to you. Without your money, they don't make any money, because their deal doesn't happen. So they are motivated to make any deal look as compelling as possible—just as any normal, prudent person would when selling a product. And just like in other industries, in real estate the sellers (i.e. the developers) are wholly dependent on buyers (i.e. investors) for their livelihoods. Without capital, no real estate projects would ever be developed.

Developers, therefore, understand how vitally important it is that they present projects that appeal to investors, because, if they don't, investors won't invest. And as in the world outside real estate, sometimes what is demanded isn't necessarily in the best interest of those consuming—think of fast food or tobacco, for example. In these cases, the seller will still sell a product to a buyer provided (i) it's legal and (ii) the buyer continues to demand the product. In a zero-sum game like real estate, this can compound the principal agent conflict for developers by creating situations where they become tempted to develop something not because it is in their best interest, but because it is demanded by investors.

In other words, morally flexible developers might migrate from building something in their best interest because it makes sound financial sense to building something that investors want but that doesn't necessarily make sound financial sense. In those cases, to protect themselves, the developer may layer fees on to a project, include waivers in the contract, and otherwise structure the deal so that they make money no matter how the deal performs. Once that happens, developers and investors are at odds with each other, yet they continue to court each other for the financial benefits they anticipate from working together.

As I mentioned earlier, Robert Ringer says in his book that a developer will build an apartment building in the Sinai desert if someone gives him the money to do it. In other words, some developers will take on inherently bad deals *if* someone pays them to. If an investor says, "I'll give you all the money you need to build an apartment in the middle of the desert that nobody needs," the developer will say: "Yeah, fine. Sign me up. If you're gonna pay for it, I'll do it."

Of course, that's not true for all developers. Your job is to pick out the sponsors who don't succumb to doing business this way.

How risk profiles have changed

Continuing the theme of investor-led deal structuring, while Ringer was writing in the 1970s, the advent of online real estate investing has led to a new version of this phenomenon. One example of this is how the preferred return is treated and paid.[1] Historically, meaning pre 2012, a real estate project would only pay out to investors when it had enough cash flow to be able to make distributions. This cash flow would come from either positive NOI or a liquidity event of some sort, like a sale or a refinance. In some cases, this has changed; and here's how.

The old way of doing things

Before the JOBS Act, it was simple: You invested in a deal in return for a preferred return—interest at a given percentage rate per annum—and a share of the profits in the deal. However, if you put money into a deal that paid 8 percent interest, but the deal had no cash flow, the sponsor couldn't pay you. Instead, that 8 percent owed to you would accrue, building up until the deal could pay you. You might have earned 8 percent per year over four years, but payment would not have been made until the building was refinanced or sold. That is the traditional structure, and the one that makes the most intuitive and economic sense.

It was the accepted structure for real estate projects, because sponsors had preexisting relationships with their investors. They met them in person and investors accepted the logic that they would not receive a payment until there was sufficient cash available in a project.

The new way of doing things

A new way has evolved that, though seemingly attractive, isn't necessarily in anyone's best interest. For instance, if the deal has no cash to pay investors for some reason—maybe it's a new build or the sponsor plans to gut an apartment building—there is pressure on sponsors to pay interest current (not accrued). This is because 'the crowd' does not behave in the same way as individuals.

The way things evolved post JOBS Act was that the first to take advantage of the deregulation of general solicitation were not sponsors directly, but those who created marketplaces to finance real estate deals—the crowdfunding websites. These websites found themselves listing multiple deals at the same time, giving investors, for the first time in generations, the opportunity to compare one deal against another, side by side on the same page.

The most prominent differentiating factor between one deal and another is how much they pay to investors. Sure, there's a lot more to it than that; but when faced with a plethora of investment opportunity, the most natural thing to look at first is which (on the face of it) seems the most lucrative. One of the ways that the first deals placed on crowdfunding websites competed with each other was via the level of preferred returns they offered. Sponsors started to compete by offering higher preferred returns, but there's a limit to how high that can go without either decimating a deal or extending credulity.

The next way sponsors started to compete was according to *when* the preferred returns would be paid. They changed their offerings from accrued (i.e. paid once a project had the cash to make such payments) to current payment (i.e. from day one). Well, who wouldn't want that! Investors became inured with the idea of current payment on preferred return, and this started to dominate the competitive landscape as a default option for sponsors to offer to investors.

Here's how it works. Imagine you have invested $10,000 in an apartment deal. Pre crowdfunding, if there was no NOI coming out of the building (perhaps because the sponsor was renovating it and had terminated rental contracts with all prior tenants), the preferred returns owed to you as an investor would accrue, and you would receive no payments until there was sufficient income in the project.

Now, even without NOI—no positive cash flow coming out of operating the building—some sponsors have become accustomed to paying investors preferred returns even when the building has no cash to pay them. And there is only one way to accomplish this magic trick of creating money from nothing to pay investors—sponsors must raise more money. And guess where that comes from: the very investors who are demanding that their preferred returns are paid current. Instead of raising, for example, $1 million in equity for a project with an 8 percent preferred return, if the sponsor believes it will take a year before the project has enough NOI to pay the preferred return, they will raise $1.08 million from investors. Put another way, investors end up being paid a preferred return *with their own money*.

Because a sponsor can trumpet the benefit of paying preferred return immediately, this looks good to ill-informed investors. But it adds extra layers of risk to a deal. The more debt you take on and the more equity you pile in that is earning preferred returns, the greater a burden it adds to the project.

It's unstable and illogical, but you will see this in some deals online. There are various gradations of the same theme, and all of them are classic cases of the tail wagging the dog—the investor dictating terms that are not in their best interest. In the age of real estate crowdfunding, sponsors are still putting deals together reminiscent of Robert Ringer's desert apartment buildings, because there'll always be a sponsor willing to figure out a way of accommodating a deal even if it is illogical to build.

Higher returns = higher risk

Another example of the tail wagging the dog is investors who flock to higher-return deals while forgetting one simple fact: the higher the returns, the higher the risk.

I've spoken to many leaders in the real estate crowdfunding industry—founders and CEOs of most of the major marketplaces—about this phenomenon, and here is what they say: the higher a sponsor projects the returns on his project are going to be, the more popular it is with investors. It's a direct positive correlation. In the same way that sponsors have learned to attract investors by offering current pay on preferred returns, they have learned that to attract the most investors in the crowdfunding world, they must offer the highest overall returns.

Again, investors are driving this; yet it can be hazardous to their financial health. Of course, all things being equal, investing in something with a higher return makes sense—but in real estate, not all things are necessarily equal.

Where there is a rush to attract investors by competing with headline IRRs, there is a danger that this will motivate sponsors to 'stretch' underwriting assumptions to deliver higher headline IRRs to attract the unwary investor.

If a sponsor is shooting so high—perhaps unrealistically high—what happens if they miss the mark and fall short of the target? An unsophisticated investor might figure that because the project appears so lucrative and profitable, if the sponsor falls short, then they'll still make a profit. Better, the thinking goes, to aim high and fall short than to aim low and fall even shorter. But there is a major problem with that line of thought: The logic is totally wrong.

In real estate, *the more unrealistic the sponsor is on their projections, the higher the chance that you will lose everything.* If the sponsor seems too optimistic, you as the investor have a much higher chance not of making less than projected but of losing everything. It isn't as simple as falling a bit short of the projection. More likely, there is a higher chance that the deal will totally fail.

Assessing risk

Real estate investing is inherently a high-risk occupation, and when investors drive the decisions professional sponsors are making, it can become quite hazardous. I've personally worked on salvaging huge multibillion-dollar portfolios of real estate investments that had all gone bad, and I have directly handled over a billion dollars of distressed real estate for both banks and private equity funds. I've seen every type of real estate that there is, and I've seen every possible way that deals can go wrong. In fact, I've learned a lot more from watching how things can go wrong than from seeing how things go right.

Even sophisticated, seasoned sponsors can get caught up in macro events completely beyond their control. In the rest of this chapter, instead of concentrating on doing real estate the right way, we'll walk through what can go wrong. Along the way, I'll describe some common real estate errors and show you how to reduce your risk of losing money.

Learning to do it right by seeing others do it wrong

A fabulous way to learn how to do something right is to watch somebody else do it wrong. There is one side effect of that: You could become very conservative in your investment practices (though there's nothing wrong with that). There's no need to get into a high-risk investment environment like commercial real estate in a cavalier way.

The good news is that you don't need a billion dollars' worth of distressed real estate experience to be able to understand this. You can figure out how things go wrong and learn how to manage the risk that things will go wrong. Although you'll never totally eliminate that risk—there's always risk—your goal should be to manage it.

The nice thing about real estate investing is that it is required by law that sponsors disclose to you every way a deal can go wrong. They must lay it out for you in a section of their offering documents (the ones you must sign to participate in the deal). This is the 'risk factors' section. It is a great resource for you to learn from.

What sponsors MUST tell you

When a sponsor explains a deal to you, they are required by securities laws to disclose the risks inherent in the deal. At the same time, that person is trying to encourage you to invest in the deal. When someone is raising money for a deal, they want to present it in the best possible light. Yet they are mandated to explain the ways that it can go wrong; to explain the various ways you could lose everything you invest. Most don't tend to do this overtly. After all, what seller wants to tell the buyer everything that could go wrong. If you were a sponsor raising money, why would you be overt with the bad news? But sponsors must explain the risks, and typically they do this by burying the risk section somewhere deep in huge contracts.

One thing to keep an eye out for is sponsors and platforms who are up front in discussing risks in detail, whether in educational materials on their websites or on social media. Those sponsors who are transparent about the risks of real estate investing— particularly those who address macroeconomic issues like recession, the impact of interest rate fluctuations, supply and demand, entitlement risk, etc.—may be sponsors worthy of greater attention. They are less likely to succumb to the temptations of pandering to imprudent investor demands and more likely to be conservative in their underwriting. Although their headline returns may not be as compelling as others, those who are more open about the risks of investing may stand a higher chance of delivering on their more modest projections.[2]

In my consulting business, where I advise real estate sponsors on how to raise equity capital online, I always say that the first thing they should discuss in their communication with prospective investors is the risk involved in investing. This is because it will enable them to establish closer relationships with investors who are in it for the long haul and who are not just looking to make a quick buck.

While responsible sponsors are unafraid to address risks head-on, many do not. A prudent investor can begin the process of learning about real estate investing from the risks section in any contract, as you will be doing in this chapter. The fact is, however, that most people don't pay enough attention to the risks. All they want to hear is the good news, because it's natural to focus on how much money you're going to make. Sellers only want to talk about how great their project is, while investors want to hear how much they're going to make. It's human nature.

I was not immune to the same fault in reasoning when I first invested in real estate in the mid-1980s. I vividly recall having been attracted to a multifamily ground up development company's proposals. During the in-person pitch meetings that I attended, I recall being hyper-focused on how much money I was going to make, what the preferred returns were, and how many multiples of my investment I was going to be making. I also remember being overwhelmed by the enormity of the offering document that I was given to sign prior to making my investment.

I figured that the only area that really mattered in the documents was the section that talked about how much I was going to make. I went straight to that section to make sure that the contract included terms that reflected exactly what I had been told I was going to be paid. I didn't give any thought to any other section of the contract, believing that as long as the payment terms were accurate, I would be alright. In the end, I lost everything and found myself not only with no recourse to help but also in debt because I had foolishly borrowed against the money I thought I was going to be making in these deals.

Today I am a member of various investor groups, some of whose members invest in crowdfunded real estate deals online. When RealtyShares collapsed, those group members who had invested in deals on that platform announced that they were reading the offering documents for the first time as they tried to figure out what rights they had— and discovering to their chagrin that they had none. I found it astonishing as I read their comments that they had not paid any attention to the offering documents when they signed them and had only started to look at them in detail after it was too late.[3]

Not only do the offering documents for any real estate deal define everything you need to know about your rights and responsibilities as an investor; they also provide a road map for understanding real estate investing and are a great place to start to unwrap the risks of investing in real estate. You can use the risk section to learn about the risks, and you can also use it to find the best deals and identify the best sponsors. In short, the risks section of any set of offering documents can be a great learning tool, and we are going to go through how to take advantage of that now.

CYA[4]

Regulations mandate that sponsors disclose risks in advance, but sponsors have also been advised over the years that it's better to disclose risks up front, because it will offer some legal liability protection for them if things go wrong. If a sponsor pitches that a deal will be a great success, an investor might complain if that deal actually fails; in this scenario, the sponsor can point to a disclosure made in a single sentence, buried somewhere in a 200-page contract, that the cause of the project's failure was one that the investor had acknowledged as an acceptable risk. Including an explanation of the risks of a deal in the contracts reduces sponsor liability down the road, which mitigates the risk of being successfully sued by a disgruntled investor.

That doesn't mean that sponsors cannot be sued. Everyone can be sued. In fact, there's no way of preventing yourself from being sued. Anyone can sue you for anything. It happens. But having these risks in the contract lowers the chance of an investor prevailing in the lawsuit.

What risks are sponsors required to disclose?

The sponsor almost always discloses the ways that a deal might fail. They disclose how they might be putting their 'thumbs on the scale' to skew things in their favor. Disclosure is, for example, a great way to eliminate conflicts of interest. If a sponsor discloses a conflict of interest, they can say that investors were prewarned and if they still buy into the deal, at least they knew there was a conflict and went in with full knowledge. They tell you everything, but they also bury details out of sight, deep in the contract. Therefore, the deal is weighted in their favor and not yours, because you have to find the relevant information within the documents.

When you consider a deal, if you only look at the glossy offering memorandum with all the nice pictures, all you get is the sales pitch. Drilling down on the small print will help to reduce your risk because you will learn:

- how to identify the risks involved in real estate investing;
- what you can learn from those risks about the development process; and
- how to protect yourself from the downsides.

As I've discussed, a great starting place is the section on risk factors in the offering documents of any deal. Those sections don't have pictures or promises of a big payday, but they reveal everything that can go wrong and give you a blueprint for what to look for in a deal.

No free lunches

Real estate crowdfunding is not free. You are using real, hard-earned money to invest in a deal. Whether you're an investor or a developer, neither a lender nor a borrower be without reading the small print. By walking through the small print, you learn how to properly underwrite a deal. If you don't want to charge into a deal with your eyes wide shut, but instead want to find the best deals and the best sponsors, it requires looking at and properly interpreting the risk factors.

A case study in risk factors

This section provides an overview of a deal I worked on some years ago, one of my own sponsored projects. One reason for choosing my own contract is that, at only 2,038 words, the risk section I wrote was shorter than many of those you're likely to find online. That's practically bite-sized compared to some you'll find online. I've seen some that run to almost 12,000 words and cover 20 pages or more.

Below are selections from my own offering documents' risk factors section. The deal I was working on at the time was one of the largest small-lot subdivisions ever to have been developed in the Los Angeles area. It was an 18-home subdivision in an infill location in the Echo Park neighborhood of Los Angeles, close to downtown. The project was a for-sale housing project, with a worst-case scenario that I was going to rent the units if the for-sale market declined for any reason. In the end, it was one of the projects that I was fortunate enough to sell in 2007, concerned as I was at that time that the market was going to collapse—which, as you know, it did the following year.

Below I present the various risk sections, review key terms, and offer ways of mitigating the various risks. These verbatim highlights are taken straight from the risk factors section of my offering documents. I've simply highlighted certain parts of the risk factors section for you, providing the exact wording followed by an analysis.

Real estate development risks

> *The business of the Company will involve the development and likely construction of the several contiguous properties into multiple single-family residences ("Business"). Development projects are inherently high risk with exposure for losses ranging from poor construction, disputes with vendors and contractors, delays due to inspections, contractors' or subcontractors' poor or slow performance to additional expenses resulting from unforeseen issues with the Property that could result in you losing all your investment.*

This first section tells the investor that real estate development projects are inherently speculative and carry significant risk, and that investors could lose all of the money they invest. It continues:

> *Real property investments are subject to varying degrees of risk. Revenues and the value of the properties may be adversely affected by the general economic climate, the local economic climate and local real estate conditions, including the perceptions of prospective tenants of the attractiveness of the properties; the ability of the Company to provide adequate management, maintenance and insurance; and effective marketing of the properties. Cash flow could be adversely affected if the Company is not able to develop and construct the housing units on the properties*

on budget or if delays in marketing and selling units occur. Real estate values may also be adversely affected by such factors as applicable laws, including tax laws, interest rate levels and the availability of financing.

By its nature, real estate is speculative. Once you invest, almost anything can happen. Un-insurable acts of God (floods, fires, etc.) can wipe out your investment. Economic conditions can kill a deal. Rising interest rates can profoundly impact a deal's success. In short, the entire deal could fail for numerous reasons. In other words, do not invest anything you cannot afford to lose *completely*! Even experienced, informed people lose money, for any number of reasons.

Environmental risks

I remember working on a deal some years ago in Los Angeles. The deal was on North Figueroa Avenue and had been brought to me by a city inspector. The property was in considerable disrepair and the city wanted someone to buy it from the present owner in order to improve the site. The owner, it turned out, was homeless and had inherited the property from his father. I met with him several times. It turned out that when he was a young boy during the 1950s, the property had been used for his father's dry-cleaning business. That was a time when dry cleaners used a highly toxic chemical, tetrachloro-ethylene, to clean clothing.

The old fellow reminisced pleasantly about how a truck would show up, pump out all the old chemical from the cleaning equipment, and then pump in fresh, new chemical. The problem was that the equipment leaked. To capture the leaking chemical, his father would put a metal bucket underneath the equipment. The son's job was to get rid of the leaking chemical, and he told me that he used to do that by going out back into the yard and in pouring the chemical on to the dirt. He enjoyed watching it seeping into the dirt and disappearing from sight.

I conducted a Phase I environmental study,[5] which confirmed that the property was and had been, in fact, a dry-cleaning business. As a result of this and in combination with the stories the owner told me, I commissioned a Phase II study that involved drilling holes into the ground all the way down to the bedrock to see if there was any contamination in the soil. When we pulled up the soil samples during the Phase II investigation, the smell of the chemical hit us. It was intensely strong and pungent, like the smell of that glue used to cement plastic.

Of course I passed on the deal, but had it not been for the Phase I and Phase II studies in combination with a little of my own research with the owner, I could have ended up buying into what had appeared to be an amazing deal that would have ended up costing me a fortune to clean up.

The section in the offering documents that addresses this sort of environmental concern might read something along these lines.

The Company has undertaken what it believes to be adequate testing of the property and is not aware of any environmental contamination. However, environmental issues with real estate will always exist. In order to mitigate any potential risks, we will obtain a Phase I environmental assessment, and dependent on the results of that assessment, in our sole and absolute discretion, may obtain additional environmental reports. We will not close the purchase on the Property unless we are generally satisfied with the environmental status of the Property.

Environmental risks include not just those from dry cleaners, but also possible contamination from asbestos, lead paint, or underground gas tanks at former gas stations. Old dry-cleaning facilities are some of the worst. What I learned about how the toxic chemicals used for dry-cleaning filter down into the ground is that if they hit the water table, they can get carried underground and, as in my Figueroa deal, create a huge plume under neighboring properties. Not only that, but you could invest in a property that has a clean Phase I assessment, showing no signs of environmental damage, but if it sits next to contaminated property, your site could also be contaminated. Contamination is a super-high liability that can come with catastrophic costs, so deals with environmental issues are ones you want to avoid. The risk is in potentially being required to remediate the previous owner's contamination. As soon as a sponsor's name goes on a title, they are on the hook for any required clean-up, even if they discover the problem and are successful in selling immediately. What's more, future owners could even come back and sue to clean up the mess.

As an investor, the way to mitigate environmental risk in a deal is to examine the steps the sponsor took and learn how deep they got into evaluating and testing the property to make sure that any necessary clean-up has already been handled.

Americans with Disabilities Act

The Americans with Disabilities Act of 1990 (the "ADA") requires all public buildings to meet certain standards for accessibility by disabled persons, and local zoning codes may also add restrictions that apply to private properties. Complying with the ADA can be costly and time-consuming to implement.

ADA compliance can add significant time and cost to a project. If a sponsor doesn't know whether their project needs to be ADA-compliant or if they failed to consider ADA compliance, and they suddenly get hit with compliance requirements—or litigation from someone unable to gain rightful access—that can add an enormous cost to a project. For example, if there's a new ADA law and a property transfers to new owners, the new owner is generally responsible for implementing those changes.

In many cases, if the sponsor has acquired an older building to renovate and new ADA laws come out, or came out subsequent to the original construction of the building, the building is exempt, having been grandfathered into the older regulations. This likely changes, however, as soon as the sponsor files for permits to make changes to the property. As soon as they need a permit for their renovation, the new laws spring into force, and the sponsor may be required to make the entire development ADA-compliant.

As an investor, you can mitigate this risk by making sure that the sponsor has checked for ADA requirements and has accounted for any costs of meeting regulations.

Zoning and entitlements

The Project is subject to extensive building and zoning ordinances and codes enforced by the City of Los Angeles' planning department, which can change at any time.

Zoning is how a municipality defines land use within its boundaries. The city determines that certain areas must be used only for commercial purposes, for example—or residential, industrial, parkland, or recreational use, etc. When a sponsor buys a property, you want to be sure that they know which development type the city allows on that property. Buying a piece of property to build apartments is no good if it's zoned for industrial use, because the city will simply not permit a residential use in that area.

There are two ways that zoning can negatively influence a development. The first is that a location may not be zoned for the proposed use. This will require the developer to apply for a variance from the existing zoning, which can be an onerous proposition with no guarantee of success. The second is that zoning decisions often come with discretionary layers. What this means is that the local politicians and neighbors can weigh in with their opinions on whether or not the proposed use is acceptable. This kind of discretionary process has a very high risk, because a sponsor often has little control over the process and the proposed project may not ever be approved by the local jurisdiction.

Consequently, adhering to local zoning codes can add significant time and money to a deal. It can be so difficult getting the proper permissions to build, develop, or redevelop a site that without proper planning a deal can be completely stymied.

To ensure that this risk has been properly mitigated, you have to check that the sponsor did the work and drilled down on the zoning requirements in detail. Did they find out what the regulations are? Do they know what the city wants? Have they spoken to planning department staff, won the support of local politicians, and engaged neighborhood groups in dialogue?

Another way to approach zoning is to find out the project's stage in the permitting process. If the sponsor has already taken their deal through the zoning approval process with the city, then the entitlement risk is mitigated. Because getting entitlements can be such a huge risk, it is crucial that you know exactly what stage the project is at and that the sponsor has completed all entitlement requirements.

Incidentally, getting entitlements can itself be a valid development strategy. My small-lot subdivision project in Los Angeles ended up being a very lucrative deal, because I had taken the project all the way through the entire entitlement process and sold it as an entitled project. In fact, some developers never actively develop real estate; rather, they specialize in getting entitlements and in selling to someone else who will develop.

Insurance[6]

> To mitigate casualty and other risks from materially and adversely affecting the project, the Company will carry comprehensive liability, fire, extended coverage and rent loss insurance with respect to the properties, with policy specifications and insured limits customary for similar properties. There are, however, certain types of losses (generally of a catastrophic nature, such as wars, terrorism, floods, landslides or earthquakes) which may be either uninsurable or not economically insurable. Southern California is a high-risk geographical area for earthquakes and landslides on hillsides. Depending upon its magnitude, an earthquake or landslide or other subsidence could severely damage the properties, which would adversely affect rental income and other expenses. The Company will not likely maintain earthquake or flood insurance for properties, unless such insurance is required by lenders. An insured loss may exceed available coverage under applicable policies of insurance. Should an uninsured or under-insured loss occur with respect to the properties, the Company could lose both its invested capital in, and anticipated profits from, those properties. In the case of an uninsured casualty on properties with debt which is recourse to the Company, that Company would remain obligated for any mortgage debt or other financial obligations related to the properties. The Company may procure and maintain insurance through blanket policies of insurance.

Sponsors need to buy insurance. When looking at a deal as an investor, you want to know what types of insurance the sponsor carries along with policy limits, coverages, and exclusions, including the types of casualties each policy covers. Part of your due diligence is making sure they have coverage for loss up to a sensible limit.

Some things are not insurable, and the term 'material adverse' is one to remember. The idea that material adverse changes had occurred was used by some banks during the 2007–2008 financial crisis to justify terminating construction loan agreements. Typically, construction loans are advanced as a developer completes the various phases of construction successfully. Inside the loan agreements were exemptions that allowed banks to withhold further lending if they determined that material adverse conditions had impacted the project.

During the recession, some banks determined that the recession itself was a material adverse condition, because property prices had dropped so much. Those banks then stopped advancing further draws on construction loans, causing borrowers (i.e. developers) to abruptly run out of the money that they thought they had access to. It was a controversial tactic used by banks, but was extremely effective in preserving their loss positions, usually resulted in a foreclosure of the property, and was completely uninsurable.

Real estate is not a liquid investment

> Any return to the Investors on their invested capital will be dependent upon the ability of the Company to develop and market the properties profitably. Such profitable operation and sale will depend, in part, upon economic factors and conditions which are beyond the Company's control. There can be no assurance that the Company will be able to operate the properties profitably. As a result, there can be no assurance that the Investors will recover their investment in the Company nor any guarantee of when.

Translation: Real estate is not cash. Cash is 100 percent liquid. Real estate transacts very slowly, making it illiquid. It may happen that the original business plan to build or renovate a property and then sell it for a profit is rendered impossible by market conditions.

There are a few ways to mitigate this particular risk. One is an approach I have seldom, if ever, seen outside of the institutional private equity world; it is to underwrite a deal based on three scenarios: best case, worst case, and most likely. There are two core benefits of this approach to underwriting a deal. One, obviously, is that seeing the best, worst, and most likely scenarios is highly illustrative of a deal's full range of possibilities. But two, perhaps more importantly for you as an investor is that it demonstrates that the sponsor has given careful consideration to the various possible outcomes and not just pitched you with the best-case scenario. As the worst-case scenario is seldom 'advertised' by sponsors, it is completely legitimate for you to ask a sponsor what is the worst-case scenario they have contemplated and how they have addressed that possibility in their planning.

One way that I have done this, for example, is to underwrite a for-sale project as a rental property and then base all borrowing against projected income from rentals. Typically, this will not yield as high as a return as the for-sale scenario, but it does create an upside in that I can actually hold on to the property during a downturn instead of losing it to foreclosure.

This requires making assumptions regarding how bad the downturn might actually be. One way to predict this is to look at the location and asset type and determine how far prices declined during the most recent recession. Assuming that in the worst-case scenario the market will go down at least as far as that, a prudent sponsor will underwrite to that level in order to set a decent floor on their assumptions. This might include assuming that rents drop to a certain level, that cap rates will rise to a certain level (i.e. that sale prices fall), or that vacancy rates will rise to a certain level—all based on historical patterns.

I have even seen ultra-conservative sponsors underwrite in their forward-looking projections a decline in rental rates alongside an increase in vacancy rates, though I have only ever seen this once. In this one instance, the sponsor assumed a negative rental rate growth in years four and five. As I looked at his proforma, I asked him why he had done this, and he told me that it was because they anticipated a recession during that period and that they expected rents to actually come down and vacancy rates to increase. In all other cases—as you are likely to see in deals you look at—when projecting into the future, sponsors almost always apply a positive inflation rate to rental income growth.

Another way you may see sponsors underwriting to truly conservative numbers is by increasing cap rates over time rather than leaving them steady, or even reducing them. Low cap rates mean higher exit prices and are used together with NOI to project exit values. Even a small reduction in cap rate predictions can lead to a sizeable increase in a building's projected value, even if NOI is projected to remain static. Conservative sponsors, especially those operating at the end of the expansion phase of a cycle, will assume that prices will come down and will underwrite to those by reducing their projected exit caps.

There are two other ways that you can mitigate the risk of market fluctuations that might cause you losses (or decreased returns) because the developer was prevented from executing on their original business plan. One is that you make sure the sponsor has solid experience investing in both the asset type and in the location where the investment is being made. The other is to see if the sponsor has addressed the possibility of a recession on the horizon in any thought leadership content posted on their website or sent to you by email. A sponsor who ignores the possibility of a recession is inevitably going to become victim to one. But one that addresses the possibility head on and who accepts that a recession is inevitable at some time is a sponsor who has a pragmatic and realistic view of real estate investing and is more likely to survive—and possibly to thrive—during a downturn.

Beyond the control of the developer

> Factors that could affect the value of the investment that are outside of the control of the Manager include, but are not limited to, interest rates fluctuations, competition from other developments, national or local economic conditions, zoning changes, environmental contamination or liabilities, uninsured losses, and undisclosed defects.

No matter how diligently a sponsor conducts their research before buying a property, especially in redevelopment or adaptive reuse projects, there is always a chance that they miss something. While in theory this is within the control of the developer, in reality some inspections are considered 'destructive inspections', where it is necessary to remove something or drill into something to discover what lies beneath. Sometimes it is not possible to carry out thorough investigative due diligence and, as a consequence—again, especially in redevelopment or adaptive reuse projects—it is almost inevitable that additional, unpredicted costs and expenses will be uncovered.

For example, asbestos can be a major risk, because it is very time-consuming and costly to remove. While the presence of asbestos can be determined prior to acquisition, it is not always possible to find out how much of it there is. If the developer wants to remove walls in order to install some bathrooms but suddenly finds asbestos, that's a major issue. Indeed, in big enough projects, removing asbestos can cost millions of dollars.

Inability to sell as projected

> There are numerous properties which will compete with the properties. Some of the competing properties may be better located or owned by parties better capitalized than the Company. Although ownership of competing properties is currently diversified among many different types of owners, from publicly traded companies and institutional investors to small enterprises and individuals, and no one or group of owners currently dominate or significantly influence the market, consolidation of owners could create efficiencies and marketing advantages for the consolidated group which could adversely affect operations of the properties.

This is rather a long-winded way of saying that there is competition in the market that is out of the control of the sponsor. There are various ways that this can take effect. In the particular contract that this risk was quoted from, the business plan was to develop land for single-family homes for sale in an infill location of Los Angeles. Over the course of the 18 months to two years it was going to take to develop the property, competition could spring up that would pose a threat to the ability of the sponsor to sell the homes as planned. This might involve the emergence of competing brand-new for-sale developments. One way to assess this risk would be to check with the local planning department to see what projects have been submitted for approval. Of course this doesn't completely mitigate the risk of new projects coming online, but understanding that the entitlement process would take many months to complete, any projects commencing the this process at a later date would, logically, not come on the market until after the subject property had been sold.

Provided, therefore, that the sponsor had properly assessed the pipeline for new product and concluded that there would be no impact, the subject project would have a competitive advantage sufficient to rely upon the underlying exit assumptions.

Another risk for a for-sale housing project in an infill location is that an apartment building might be converted to for-sale condominiums. Similarly, this would require a lengthy entitlement process. But by understanding what is in the pipeline at the local planning department, it is possible to assess and predict with a certain degree of reliability what competing projects might come on stream.

I have worked on projects where the buyer has made it a condition of the closing of a sale that a piece of developable land has a complete absence of any competing projects being submitted to the planning department. One project that I worked on, for example, was a senior housing project where the buyer had determined that there was only enough room in the market for one such project. They made it a condition of the sale of the land that no other senior housing projects would be in planning at the time the sale was closed. They figured that as long as they were first to market, they would capture the bulk of the demand for senior housing and find themselves with a successful project that itself would be a barrier to entry for other similar projects.

Development and construction risk

A significant risk associated with new development is the construction process itself. In fact, a lot of sponsors don't want to buy a piece of land and build from the ground up.

There are all sorts of risks involved in that sort of development. The sponsor might underestimate the time it will take to build or miss something in the estimated cost of construction. Subcontractors may not show up or costs could increase over time, increasing the necessary budget. The architectural drawings could have flaws that aren't uncovered until construction begins in the field.

I worked on a project some years ago when I had first returned to the United States from Japan and was doing some custom homebuilding for clients. One of my clients was a celebrity who had gone to great expense to hire one of the top architects in the country to design his gorgeous hillside home. The architects had called for some I-beams in order to create high ceilings and large internal spans. On one wall, he had also planned for a fireplace to sit in between two very large doors, that would open to a gorgeous garden.

The doors and fireplace had been designed to be standard sizes in order to keep the cost down. Unfortunately, it was not until we were building the house that we discovered that the architect had called for two I-beams, one on each corner of this wall, but had not accounted for the size of the beams themselves. Consequently, we found ourselves with two 12-inch I-beams that were sitting in 4-inch spaces. Put another way, the inside dimensions of the wall were 16 inches too small, 8 inches at each corner.

This meant that we could not use standard doors or fireplace and would have to order and purchase custom materials that would take a long time to manufacture and deliver, at a considerable increase in cost. To mitigate this, we relocated the fireplace, because, although architecturally it was not what the owner had originally wanted, it did allow him to remain closer to his budget and to keep us on schedule for the construction.

While it may seem something of a generalization, you can, with reasonable confidence, assume that a project will take longer to complete than the sponsor projects, and will cost more. Any sponsor who tells you otherwise is either inexperienced, naïve, or can demonstrate from a history of perfectly executed projects that they are among the very rare developers in the industry to finish on time and on budget.

The only time I have ever seen this rule broken was when I was building huge entertainment complexes in Japan for Universal Studios. Despite the enormous scale of the projects I was developing, the general contractors who I had hired to do the construction guaranteed that they would finish on a specific date. In fact, so reliable were these general contractors that 18 months ahead of our official openings I was able to book media spend (my marketing and advertising) to commence on a specific date. Booking so far in advance helped us to reduce our media budget, but gave my board of directors heartburn, believing, as they did, that it was not possible to be so precise with the timing of such large construction projects.

What gave me the confidence to trust that we would open on time was the general contractors' explanation of how they would complete on schedule. They told me that in the event they started to fall behind schedule, double or even triple shifts of workers would be put on a job to ensure that the task finished on time.

At minimum, a sponsor needs to have completed a very detailed line item estimate of all the costs involved in a proposed construction. Ideally, they will have taken three bids for each line item and balanced the resulting estimates with a subcontractor's reputation for reliability. There is little point hiring the lowest bidder if they don't show up on time or deliver shoddy work.

You should also ask to see the developer's timeline, which will outline the critical path to completing the project and show when each part of the process and trade is scheduled to start and finish.

Another important thing to know when mitigating construction risk is who pays if a project goes over budget. For example, in a deal projected to cost $25 million, investors contribute $8 million and the sponsor borrows $17 million from the bank; but if there's a cost overrun so that it will now cost $27 million, who pays for the overage?

While you can't mitigate cost overruns, you can mitigate the risk that you'll need to put in more money. You do this by carefully reviewing the contracts and finding out exactly what the mechanism is for addressing cost overruns. Also, you could simply ask the developer how often they have run over budget and how they handled that.

In many cases, you will find that the investors will have the option but not the obligation to contribute any additional capital that a project needs due to cost overruns. This additional capital can be brought into the project in two ways: as debt or equity. Either option has its pros and cons, but whichever it is, you need to be sure to read the contracts and ask the sponsor how they handle cost overruns and make sure that you are comfortable with those terms. Here's how those kinds of risks are articulated in the sample contract we are looking at here.

> *Prospective Investors should also be aware that, if the properties experience operating deficiencies, the Members may find it prudent or necessary to provide any required funds to meet such deficiencies and protect their investment in that Company. No Investor will have an obligation to contribute additional capital to the Company beyond the amount stated in its respective Subscription Agreement. However, if the Manager determines that the Company requires funds in excess of those to be provided by this Offering and/or the Company's operations, the Manager may seek to obtain such funds through outside financing (including loans by the Manager and its affiliates) and/or through the sale and/or refinancing of properties or the sale of additional Units that may result in dilution of an Investor's ownership interest in the Company if an Investor fails to pay its proportionate share.*

You certainly do not want to be obligated to make up any equity shortfalls, or be penalized for not doing so. Keep in mind that any debt that comes in will likely be at a higher priority than your equity, which will increase your risk of loss, and that any equity that is brought in will in all likelihood be superior to your equity or will be dilutive of your ownership in the project. Sponsors I have worked with to assist in raising capital have reported to me that typically it is they who make up any overages in the form of loans to the project. They take the long-term perspective, wishing to preserve their reputation with investors. But it is not always possible to make up shortfalls in this way, no matter how well intentioned, so you want to be sure you know what a sponsor's contingency plans are.

Reliance on experience of management

> *The Company will depend upon the experience and expertise of the Manager in, among other things, the selection, purchase, management, leasing, development, redevelopment, repositioning, operation and sale of the properties. If the Manager cannot manage the properties for any reason, experienced management that understands the intricacies of the foregoing issues may not be readily available, and revenues from the properties could be negatively affected. The Company cannot assure that the Manager will be able to successfully manage the properties.*

Good management is key to any project's success, and there is always going to be a key person running things: the manager. But what happens if they get hit by a bus? How do you deal with that? To mitigate this risk, you want to see that the management team has some depth and that in the event that the key man becomes incapacitated for any reason, there is a team in place that can complete the project effectively. And you probably want to stay away from a one-man shop unless they have a really compelling succession plan in place.

Keep this in mind also: as a small investor in a large project, which you will be as an investor in a crowdfunded deal, you will likely never have the right to oust management, no matter what you think of them or how badly they perform. In the pre-crowdfunding era, investors had considerably more rights than they do now, primarily because in many cases offering document contracts were negotiable and investors making larger investments in terms of the percentage of the capital needed carried greater weight in negotiations. If the manager was negligent, or if a deal went wrong, or if the sponsor stopped answering calls or delivering reports as agreed, investors had recourse that could include removal of the manager.

In crowdfunded deals, contracts are 'take it or leave it', so you want to pay particular attention to the extent to which a sponsor demonstrates their integrity by looking closely at the offering documents and making sure that there are no egregious terms.[7]

No secondary market for your investment

> *There will be no ready market for the Units,[8] and it is unlikely that any such market for the Units will develop. The price and terms of the Units offered were arbitrarily determined by the Manager, and bear no relation to the assets, revenues, or book value of the Company, or to any other objective criteria.*

As a real estate investor, you are technically buying shares in a company, and this is why the offering documents in real estate contracts are governed by securities laws. You are not buying shares in the same way as if were you to purchase shares in a publicly traded company on the stock market. On the contrary, in many cases you are prohibited from selling your shares in a real estate project, so it is important to understand that any investment is likely to be for the long term.

That's why you should not invest any money that you might need for other purposes during the lifetime of the deal. For example, if you anticipate needing money for medical expenses, to pay university fees, for a down payment on a house, or for any other purpose, you probably should not use that money in a crowdfunded real estate deal. Of course, some projects offer greater liquidity than others, but for the most part these are long-term hold transactions where, once your money is in, it cannot come out until the project comes to the end of its natural life cycle and is sold—and this timing is generally going to be at the sole and absolute discretion of the sponsor.

Macroeconomic influences

> *Global, national or local economic recessions may negatively impact the project. Interest rates changes, our ability to access credit markets, capital market conditions, employment variations, changes in real estate values caused by macro-economic cycles, tax regulations, political conditions, wars and other crises, among other factors, are unpredictable and cannot be mitigated by either due diligence or underwriting.*

There's not much you can do about unpredictable macroeconomic stuff (other than not investing anything you can't afford to lose). One way to mitigate risk is by having a long-term perspective and another is to maintain low leverage. It is a cliché to say that the three most important things in real estate are location, location, and location. More precisely, the best way to protect against macroeconomic variations is to purchase quality buildings in good locations with high-credit tenants, use low leverage, and have a long-term perspective.

And the most important of these factors is ensuring a project does not have so much leverage on it—bank debt—that it cannot withstand a decline in revenues to the point that it cannot service its debt. Having personally handled a billion dollars' worth of distressed real estate, I can say that, with some exceptions, there's no such thing as distressed real estate. There is only distressed *finance* on real estate. And the real reason that people lose real estate is because they borrow too much and can't afford to keep paying the bank when things go down.

Indemnification

> The Manager, its affiliates, and their respective officers, shareholders, directors, employees, and agents will not be liable to the Company or any of their members. The Company will indemnify the Manager against any claims, liabilities, or damages, including attorneys' fees, incurred in connection with the performance of its duties, except where such person or entity acted in a grossly negligent or fraudulent manner. An Investor will have more limited rights of action against the Manager and such persons than would be available absent the indemnification and waiver provisions described above.

This is a crucial thing to look for when you read contracts. There could be a line saying that investors agree that they have no right to sue the manager for anything under any circumstances. Even in the event of sponsor negligence, gross negligence, simple incompetence, or plain stupidity, you can't sue them. Check the contract to see if this sort of language is in there. If it is, at least know what you're getting into.

How do you mitigate that risk of not being able to sue? I know this is starting to sound like a broken record, but you must read and understand the contracts. Understand your rights and responsibilities and check for an alignment of interest with the sponsor. Figure out how much pain they would suffer if the deal goes wrong. Learn their pain threshold and see if it the same as yours. Will it keep them on track? If they put money in and the deal goes south, will they lose too?

Seeing that there is an alignment of interests is about the only way to mitigate this risk in any measure.

Conflict of interest

> In general, conflicts may arise in the allocation of the time which the Manager is able to devote between operations of the Company and other competing projects.

Understanding conflicts of interest are very important. Management may not devote all their time to the deal you have invested in; they may have other priorities and might only spend ten minutes a week on your deal. How do you control for that scenario?

Again, make sure that there's an alignment of interest. Specifically, be sure that they have a co-invest that they could lose. You also want to see how they get paid. What kind of fees and promote do they have? How much of the back end are they getting?

Legal counsel

> *The Manager has been advised by counsel who does not represent the interests of the Members. Members are advised to consult with their own attorney regarding legal matters concerning an investment in the Company.*

Typically, sponsors spend tens of thousands, if not hundreds of thousands, of dollars on lawyers to put contracts together. But their lawyers don't represent you; they represent the sponsor. Yet, while any prudent investor would think to hire a lawyer to check over documents, if you're only investing $25,000 in a deal, it doesn't make sense to spend thousands of dollars on a lawyer.

The best thing to do is to read the contract carefully to identify deal-breakers before clicking the 'I agree' button. You can reduce the cost of hiring an attorney by picking out key things that you're not sure about and then have a lawyer go through only those clauses.[9]

Final words

The section on risk factors in any set of offering documents provides a tremendous road map for understanding the DNA not only of the deal you are looking at, but of real estate investing in general. They have been crafted over decades by lawyers based on missteps sponsors have made; and, as I said earlier, one of the best ways to learn about anything is to study what happens when it goes wrong. As well as other sections of the offering documents you must sign when investing in a deal, the risks section has grown to be lengthy and detailed. This is, in part, why so many offering documents are so incredibly long—as long, in some cases, as this entire book. One contract and one risk factor section begets the next insomuch as when a sponsor comes to producing their next offering documents, their attorneys dust off the old one, add anything that needs adding based on new information, sponsor experience, or case law, and very seldom remove anything from prior versions.

As a foundation for being an effective investor in real estate, use the risk factor section of any set of offering documents as your guide for understanding real estate development. When you invest in a real estate deal online enter deals with your eyes wide open. Read the contract, learn the risks, and know what you're investing your money into.

Questions checklist

Based on some of the risks discussed in this chapter, here's a checklist of questions you can use when looking at the offering documents of a deal or, if you can't readily find the answer there, ask the sponsor before investing.

- What is the biggest threat to the success of this project?
- What kind of insurance do you have?
- What did your Phase I environmental study uncover?

- Are there any retrofit requirements you will have to deal with for ADA compliance?
- Do you have entitlements yet?
- Do you have permits yet?
- What are the best, worst, and most likely scenarios?
- How have you stress-tested your assumptions?
- How many bids do you have for each line item on the construction budget?
- How much contingency have you built in to the construction budget?
- If you go over budget, who will pay for that?
- What projects have you developed previously that are similar to this one?
- How much debt are you taking on as a percentage of total projected cost and value?
- How much money do you have in the deal?
- What fees are you taking out?

Notes

1　See Chapter 8 for a discussion of the preferred return.
2　There's a great Calvin and Hobbes cartoon that speaks to this idea. In it, Calvin (the little kid) asks Hobbes (his friend the anthropomorphic tiger): "If you could have anything in the world right now, what would it be?" Hobbes thinks a while before replying that he would like a sandwich, to which Calvin, aghast at the modest scale exclaims, "I'd ask for a trillion billion dollars, my own space shuttle and a private continent!" In the last frame of the strip we see Hobbes noshing on a sandwich and making the point that he got his wish while Calvin, who obviously didn't get his, looks on disgruntled.
3　See Chapter 9 on contracts for more detail on what to look for and how to find it in the offering documents.
4　CYA—Cover Your A★★.
5　Phase I studies are primarily records based, checking to see what the history of the site has been since it was first subdivided and made a legal parcel. If there is a history of it ever having been used for some purpose that might cause concern from an environmental perspective, like having been used as a gas station or dry cleaners, a Phase II study is conducted. A Phase II study involves 'destructive' testing whereby samples will be taken from the soils beneath the site by drilling and removing core samples.
6　Insurance joke: Three developers are sipping piña coladas on a beach in Hawaii. One says to the other, "Why are you here? Why are you not working?" The other replies, "Oh, it's terrible … I was building this building. There was an awful fire. I got the insurance. I gave up … I'm not doing that anymore, so I retired." He says to the second guy, "What about you?" The second guy says, "Well, we had this fire system up on the top and all the water fell through and it flooded the whole building. It was terrible. It destroyed everything. So I collected the insurance money, and I'm not gonna do it anymore. I came here." The third fella says, "Oh, similar … same thing happened to me. There was a terrible earthquake. The building collapsed and I collected the insurance, and decided I wasn't gonna …" The other two interrupted him and, astounded, asked, "How'd you make an earthquake?"
7　See Chapter 9 on contracts.
8　'Units' are, in this case, ownership shares in the operating company that the developer has formed to develop the project.
9　See Chapter 9 on contracts.

8   The eight financial keys

In this section, we'll cover the eight financial keys that open the doors to real estate investing. I use the term 'keys' because you can apply what you learn here to any type of real estate deal you might be considering investing in. Apply these keys to any asset class, type of real estate, or whatever investment strategy a sponsor employs for a particular deal. With these keys, you'll better understand whether a deal is worthy of your investment.

This is how I really honed my own skills. As I mentioned earlier, a few years back, I ran a division of Universal Studios—actually, a joint venture of Universal Studios and Paramount Studios—building out their real estate portfolio in the Asia-Pacific region. I was president of the division, but my direct supervisor was an accountant who saw the world through the cells of a spreadsheet. I found this to be an extremely useful discipline for determining the accuracy of decisions regarding specific line items or phases of the development process in potential investments. We'll do the same here, using these eight key concepts to examine potential deals as though through the cells of a spreadsheet.

In order to understand all the facets of a potential investment, you need to learn these eight fundamental concepts. Even though you can find dictionary definitions of the concepts, in this book, we'll review each one in light of practical scenarios. In addition, I'll share from the 30 plus years of experience I've gained and from the billion-dollar-plus transactional experience I've had so that you can understand how to use these concepts in the field and how they impact your potential investment decisions.

In the following sections, we'll discuss each of the eight keys in detail. We will start with 'net operating income' (NOI) and then move on to 'cap rate'—a very important concept. Next, we'll look at the idea of 'returns' and see what these are in a real estate deal. After that, we'll cover the 'equity multiple' and then the 'internal rate of return' (IRR)—a little-understood concept, even amongst seasoned professionals. Following that, we'll examine the 'waterfall,' which encompasses one of the headline numbers you will see: the preferred return. Then, we'll dive deep into the idea of the 'promote,' which is, in part, how sponsors get paid. Finally, we'll examine 'leverage' in a way that you have probably never seen before.

Finance drives everything

Over the course of my career, I have learned that that there are actually only two things that you really need to grasp in order to be able to understand any business or profession: the economics of the business and the legal contracts that business uses. This applies equally to real estate. In fact, I have learned about many businesses within real estate, from how to finance and develop multifamily housing developments to how to build large-scale entertainment complexes, to the ins and outs of ground up construction and renovation projects.

Even today I learn something new from my clients every day about the asset types they are involved in, like the economics and operations of downtown skyrise office buildings, creative solutions to workforce housing projects, and even how to operate a real estate crowdfunding platform. This new knowledge complements my past experience.

And I have come to realize that with enough dedication and focus, you can learn 90 percent of everything that you need to know about a new sector of real estate (or a new business) within three months, but it will take the rest of your life to learn the remaining 10 percent.

Although there are eight phases of development in real estate, as a passive investor you only need to really understand how the financial concepts influence, and are influenced by, a project and be able to unwrap the legal relationship between yourself and the sponsor through understanding the offering documents.

Once you understand the key financial concepts, you can review all phases of a project through the lens of the deal's finances. Regardless of what kind of real estate you're considering, you only need to understand the key financial terms to be able to look critically at any deal. It doesn't even matter which of the five types of real estate you choose. The specific development strategy doesn't really matter either. If you can master the key financial concepts of a real estate project, you'll be able to understand every deal you look at, and if you can read between the lines of the contracts, you'll be better able to protect your own interests.

Financial key 1: Net operating income

NOI is perhaps the most important concept in real estate investment. It is the fuel that drives every real estate deal. In brief, NOI is the cash that comes out of a deal. It is easy to understand because it means exactly what it says, yet it's still helpful to break each word down.

Net | Operating | Income

Net is what you have left after deducting all expenses from all income—the net income from operations. So the *income* that comes after paying for *operations* is the *net operating income*.

Operations are the processes involved in running or managing a building. In building management, there is an operator. The term 'sponsor' is often used synonymously with the term 'operator.'

In general, there are two financial sides of operations—income and expenses, or costs.

- Operating income includes rents or fees paid in.
- Operating costs include staffing, utilities, and other costs related to operating the property.

The resulting profit—NOI—is calculated by subtracting operating costs from gross operating income, represented by the following equation:

$$Total\,(gross)\,income - Operating\,expenses = Profit\,(NOI)$$

Income for a building can be generated in numerous ways. The most common is, of course, from rent, but there are other ancillary revenue streams that sponsors can generate; for instance, through providing premium parking for an office building or laundry facilities in an apartment building.

Here's an example of how NOI is calculated. Assume that an apartment building generates $1 million in gross revenues per year and that it costs the owner $300,000 per year in expenses to operate the building. The NOI in this example is $700,000.

The assumptions about how NOI will grow over time are carefully scrutinized by lenders—and should be by you too—as it drives a building's ultimate projected value. This is one of the most important indicators of the returns you can expect on your investment.

Financial key 2: Capitalization rate

The ideas and practices behind capitalization, or 'cap,' rates are among the most valuable concepts you need to understand as a commercial real estate investor. They are used in the world of commercial real estate to indicate the rate of return that one can expect on an investment property. Incorrectly calculating or misinterpreting cap rates is a major failure point for new commercial real estate investors. In fact, inefficiencies within the cap rate process can make for fantastic opportunities or a disaster. If you want to make sure you're the beneficiary of a good deal, and not holding the short end of the stick, **pay attention**.

The cap rate is the ratio of NOI to the value of a building. It measures the expected unlevered yield for an investment. Take, for example, a building that sells for $10 million and where the NOI for that building is $500,000. The cap rate is calculated by dividing the NOI by the purchase price of the building, which, here, gives a result of 5 percent.

One way to think about cap rate is to compare it to interest earned in the bank. If you deposit $100 into your bank account and the bank pays you 1 percent interest per year, then you receive $1 at the end of the year. In this example, the capitalization rate the bank offers for your $100 is 1 percent. They give you $1 for $100 deposited.

Cap rate provides a useful measure for comparing different types of real estate and real estate investments, because it is calculated the same way no matter what you're looking at and applies equally to all types of commercial real estate.

Since the cap rate also applies to every asset type, it allows you to compare the kinds of returns you'll get in one category versus another. For example, the cap rate enables you to compare an apartment deal with a hotel deal with a senior housing deal with a warehouse deal, and so on. And although the mechanics are a little different, you can also compare returns in development deals against each other.

There is one fundamentally important rule concerning cap rates: it can only be calculated on unlevered numbers. This means you cannot include any kind of debt in a cap rate calculation. For example, a $10 million building with $500,000 in annual income has a cap rate of 5 and yields a 5 percent return.

Let's say a sponsor has $7 million of debt on the property, so only has $3 million of equity invested in the project. Now, for the sake of ease of calculation, let's also assume the bank has generously made a loan to this borrower at a 0 percent interest rate, leaving our $500,000 of NOI intact.

If we attempt to calculate a return to the sponsor in this scenario, we'll find that the cash-on-cash returns[1] are 16.7 percent—$500,000 divided by the sponsor's $3 million investment. If the sponsor is able to borrow 100 percent of the cost of the project and still yield $500,000 in NOI, his return would be, technically, infinite in percentage terms.

One way to eliminate the distorting influence debt has on a project's returns when comparing one deal against another is to remove it completely from the calculation of returns—and that is how we arrive at the cap rate.

The cap rate formula

The cap rate is the total NOI—the profit generated from operating a building—divided by the total cost of the project including the total cost of construction, or the purchase price of the building not including debt. As a formula, it looks like this:

$$Cap\,rate = \frac{Net\,operating\,income\,(profit)}{Total\,cost\,of\,the\,project}$$
$$(either\,construction\,cost\,or\,purchase\,price)$$

For example, if you purchase an apartment building for $1 million, and the NOI is $50,000 after expenses, the cap rate is $50,000 divided by $1,000,000, or 5 percent.

$$\frac{\$50,000}{\$1,000,000} = 5\%$$

Later, we'll review some more complicated examples, but for now, let's return to the bank deposit scenario. This would be the same as if you deposited $100 in the bank, and the bank paid you 5 percent interest ($5)—though, of course, unlike interest payments from a bank, payments are not guaranteed in real estate investment.

However, even though the cap rate is probably one of the most commonly used terms in real estate, it is also one of the most frequently miscalculated items. As I have mentioned, here is no central school of real estate, and while there are correct definitions of these concepts, there is no requirement for everybody to learn and understand them in a standardized way. Consequently, they are perceived or understood differently by different people. Everyone understands real estate through the lens of their own experience, based on who taught them and how they learned about a particular concept. The result is inconsistency. Remember: at any time, if you're not absolutely sure of the meaning of a term a sponsor is using, ask them to explain their understanding of it. You might find the answer to be enlightening for a multitude of reasons.

One potential consequence of nebulous understandings of cap rates, or other related concepts, is that if somebody doesn't really understand what a term means, it can lead to errors. And errors can be exploited.

For example, a sponsor might discover a deal in which a seller has calculated a cap rate inaccurately. The seller uses the cap rate that he believes to be consistent with the market and sells the property at that cap rate. But what if they've made a mistake? They may have underpriced the asset. Consequently, the buyer—the sponsor you're investing in—may have found a bargain based on a miscalculated cap rate. They will be able to unlock value that the seller didn't know existed, because they didn't price the property correctly.

Calculating value

Cap rate can also be used to calculate purchase price by dividing the NOI by the cap rate and then multiplying by 100, as shown below.

$$\frac{Net\,operating\,income}{Cap\,rate} \times 100 = Property\,value\,(or\,price)$$

Taking the example used at the beginning of this chapter, we get this calculation:

$$\frac{\$500,000}{5} \times 100 = £10,000,000$$

You'll notice that the higher the cap rate, the lower the price of a property, and vice versa. Here's how that works. Let's return to the banking analogy to explain this idea. Let's say that we deposit $100 into our bank at 1 percent interest. In order to earn $1 a year at the bank, you need to put $100 into the bank. It's the same as a 1 cap investment: In order to earn $1, you have to spend $100 on a deal to earn that income.

Now, let's assume the bank is offering you a 5 percent interest rate. In this case, you only need to deposit $20 in order to earn $1 of interest. Thinking about that from a real estate investment perspective, that means if the cap rate is higher, the price of the property is lower. In order to earn $1 million of income per year, you only need to invest $20 million to buy a building in a market where properties are trading at 5 caps. But in a 1 cap market, to earn $1 million of income, you would need to invest $100 million in a deal.

Figure 8.1 displays cap rate data across a range of different asset types—office, industrial, retail, multifamily, and hotel—for quarter 1, 2010, through quarter 2, 2016.

The graph displays a steady trend over time, pointing to the market's overall performance. The cap rates compressed (decreased) over time. In 2010 every asset type was commanding between 9 and 10 percent cap rates, but in the six years leading to 2016, the cap rate for each type dropped to around 7 percent. Although there were exceptions and peaks and troughs in the pricing, overall, there was a downward trend in cap rates over the six-year period shown. This translates to an upward trend in prices. As you can see, cap rates are very useful, not only for comparing asset types (as in this chart) but also for identifying overall trends in the market, irrespective of the type.

As a real estate investor, the implication for you is that you can spread your investments across multiple asset types to blend returns and consequently mitigate risk. For example, you can invest in higher cap rate asset types or locations and diversify by also investing in lower cap-rate asset types or locations.

Comparing assets with cap rate

You can compare asset types according to their relative cap rates, and you can also compare different markets in the same way. For instance, you may find a ten-unit building in one market for $1 million, while in another market the price might be $5 million for the same type of building.

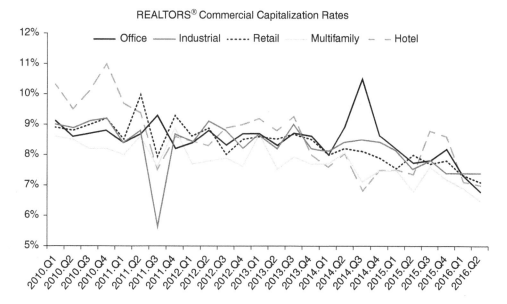

Figure 8.1 National Association of Realtors cap rates

What drives this difference in nominal dollars in two markets is the net income a building can generate. While investors are buying real estate, in reality they are also buying an income stream; and it is based on that income stream that the cap rate is used to value an asset.

If rents are higher in one location over another and costs as a percentage of gross revenue remain the same (as it generally does, with some minor variation), then higher rents, either per unit or per square foot, will yield a higher NOI, which will drive a higher overall property value.

Therefore, as an investor, you need to have a tool you can use to measure the relative value of one investment over another where income streams differ, which also lets you compare similar properties in different markets. One of the tools you can use to do this is the cap rate.

Cap rate and your investment strategy

Cap rates are also used to develop an overall investment strategy; for example, by answering the question of whether it is better to buy cheaper properties in expensive areas or more expensive properties in cheaper areas. Put another way, while it's easy to say location is the most important qualifier in making an investment decision (the cliché being that the three most important rules to investing in real estate are location, location, location), not all locations are created equal, and it is not always easy to compare one against another.

To answer this question, I spoke with Dr. Greg MacKinnon of the Pension Real Estate Association in my first podcast series. He's an economist at the association who discovered some interesting things from his research.[2]

Dr. MacKinnon conducted a detailed analysis of primary data covering a period of 20 plus years to examine whether it's better to invest in expensive areas or cheaper areas—as measured by cap rates and relative income levels in high-cap versus low-cap areas, and low-cap versus high-cap buildings.

The results of his study were that, nationwide, investments in less expensive markets (i.e. those with higher cap rates) consistently outperform, on a risk-adjusted basis, investments made in properties of similar quality in expensive cities (i.e. those with lower cap rates). These findings are consistent with the 'value investing' thesis famously employed by Warren Buffett, inspired as he was by Benjamin Graham's book *The Intelligent Investor.* What Dr. MacKinnon's findings suggest is that value investing applies to real estate as much as it does to other asset classes, like stocks, bonds, and commodities.

Cap rate as a multiple of NOI

You can also use the cap rate to derive projected added value from planned improvements by looking at it as a multiple of NOI. Let's assume that there's a building on the market with an NOI of $50,000. You purchase this building for $1 million. This is a 5 percent cap rate, but it is also a 20 times multiple of the NOI for that building. The NOI is $50,000, and you paid 20 times that ($1 million).[3]

Now think about this: For every dollar of NOI that you add, you increase the building value by 20 times that same increase in NOI. For example, if you pay a 5 cap in any market for an apartment building, any dollar you add to the rental stream that comes from it will be worth an additional $20 in the value of the building. The building went up by a multiple of 20 of the newly added NOI.

I love this idea, and here's why. If you have a ten-unit apartment building and each unit pays $1,000 per month for rent, that's $120,000 of gross income (before expenses). After deducting your operating costs of 30 percent, or $36,000, you have an NOI of $84,000.

If you want a 5 percent return on your unlevered investment, then in order to buy that $84,000 of income stream, you would have to pay $1.68 million to get your 5 cap return. Here's the calculation:

$$\frac{(\$84,000)}{5} \times 100 = \$1,680,000$$

Another way of calculating that is to say $84,000 \times 100/5$. This is your 20× multiple. It's the same idea—20 times the NOI, which is the same as saying $84,000 \times 20 = \$1.68$ million.

Now, let's say you can increase the rent by $100 per unit, and assume that this $100 drops straight to the bottom line—there are no additional costs associated with the rent increase, so it goes straight to profit. What does that mean in terms of the property's value?

It means that you boosted the building's NOI by $12,000 per year—ten units increased in rent by $100 per month for a year. Before, the total NOI was $84,000 on a property valued at $1.68 million. Now, the total NOI is $96,000, but the market value for the property is still a 5 cap. Now, that 5 cap means that the value of the property is:

$$\left(\frac{\$96,000}{5}\right) \times 100 = \$1,920,000$$

The value of the property has increased to $1.92 million. The other way to calculate this is to multiply $96,000 by 20 (the multiple) to arrive at the same result—$1.92 million. Due to the rent increase, the property value increased by nearly a quarter of a million dollars, from $1.68 million to $1.92 million.

In this example, you added $240,000 in value by merely increasing the rent by $100 per unit, or a total increase in rents of $12,000. Multiplying this amount by 20 is how we calculate the increased value of the property. That's certainly a great way to add value. In fact, many deals you can invest in online use this same concept to add value.

'Build to' cap rate

The next concept, the 'build to' cap rate, is a little bit more advanced. This is used to determine if a deal is a go or a no-go. In our example here, if a sponsor wants to buy a parcel of land for $3 million to build apartments costing $17 million, the total cost of the proposed project would be $20 million.

Scenario 1

In one scenario, once the building is completed and stabilized, the sponsor projects that the property will give $1 million in NOI. This means that if the local market is a 5 cap—people would be willing to pay to earn a five-percent return—the value of this building with a $1 million NOI would be $20 million. However, the cost of the building is, in this example, also $20 million.

Is this a doable deal? No, it's not. Obviously, you don't want to build and not make any money. This is one example of the 'build to cap' concept. The cap rate, as you now know, is NOI divided by total cost. It's a $1 million projected NOI against $20 million of construction, so the sponsor in this scenario is *building to* a 5 cap. There's no point building to a 5 cap if the market rate for properties in that area is also a 5 cap.

If you build to a cap rate that is the same as the market rate, there's no upside. You won't make any money. Plus, you have all the construction costs and market risk to bear. You could buy a building without any danger—no construction, no market risk—at the same price as it would cost you to build. All things considered, it's not worth it.

Scenario 2

In another scenario, the total project cost is also $20 million. However, in this case, the NOI at stabilization is $1.4 million, a slightly different situation. In this case, the sponsor is building to a 7 cap. The cap rate is calculated as follows: $1.4 million divided by the cost of the project (land plus construction) equals seven percent.

$$\frac{1,400,000}{20,000,000} = 7\%$$

If the market in this area is willing to pay a 5 cap, then the value of the building when he completes will be $1.4 million/5 × 100, or $28 million. The sponsor built to a 7 cap, but the market is a 5 cap. Therefore, the value is $28 million. This could very well be a 'go' scenario. The 2 percent difference between the build to cap and the market cap makes this deal look like a good deal.[4]

Variations on cap rate

Let's look at a few variations on the theme of cap rates. Apartments are often the asset type with the highest demand, which means the cap rates are among the lowest. If you're investing in apartments, you'll probably pay more for the income stream coming off an apartment than you would for any other asset type.

Apartments are generally seen as the most secure asset type during a downturn. However, this doesn't mean, of course, that you can't overpay for them. Though if you invest wisely, it can be a secure asset type, because it does have an income stream.

In contrast, land is the exact opposite. If you invest in land and get a great deal, but the market turns, guess what—there's no income from land. In fact, there's only expense in the form of insurance and property taxes. Land sucks money out of your pocket during a downturn, whereas apartments have tenants, regardless of whether they pay lower rents or the property has a higher vacancy rate than during the good times. So, in general, apartments are more desirable because they are income-producing, whereas land is not.

One issue to watch for with apartments is that at the end part of a cycle, there is usually overbuilding, especially in the luxury end of the market. As land prices go up, sponsors must buy land at higher prices to stay in business and keep developing. To justify the underwriting, they continuously increase projected rents, and the only way they can do that is by building increasingly luxurious apartment buildings and assume ever decreasing cap rates. What you find then is a glut at the high end of the apartment market during the end of a cycle with ever increasing projected sales prices, but a lack of product at the lower, more affordable end of the market.

Beware of comparing real cap rates with predictions

It is not uncommon to use predictions of rent growth alongside cap rate calculations to promote the sale of a building. Listing brochures that advertise value add apartment buildings will often use the current cap rate for the building and compare it with the market cap rate to indicate a value add opportunity for buyers.

It might say, "We're selling this building at a 5 cap on current income, but actual market rental rates are higher, and consequently, you're actually buying this property for a 6 cap." The pitch here is that if it's a 5 cap market and you're paying a 6 cap price, you're getting a bargain.

Similarly, in their pitch decks to investors, sponsors will base return projections on projections and not actual income. This idea is sometimes referred to as trailing versus forward-looking cap rates analysis, and there is a significant difference. The trailing cap rate describes the value of the building based on what it is currently earning, typically over the last 12 months, whereas the future cap rate is when the sponsor projects value based on what the building will earn in the future.

This is one of the keys to your due diligence process. Sponsors are always projecting a return, and there are no guarantees. They project what they think they can get. That's why you must be careful. The only thing you know for sure is what has happened. You can learn how this building has performed historically, but you don't know how it will perform in the future.

Final thoughts on cap rate, or read this twice

You cannot become a successful, long-term commercial real estate investor without a solid grasp of cap rates. You may luck out on one deal, or two, but eventually the market will catch up to you. The good news is that many investors do not understand cap rates, or do not show the calculation and analysis process the respect it deserves, which creates many of the best opportunities in the ever more efficient commercial property markets.

Like any investor, you want to be able to find underpriced assets. By leveraging cap rate correctly as a buyer, you can unlock incredible value. As a seller, you can avoid many of the pitfalls associated with inefficient or incorrect property valuation. Any way you cut it, this chapter is one of the most important in your journey to success in this space, so don't be afraid to give it a second or even third or fourth look.

Financial key 3: Returns

A simple way of contextualizing the concept of returns is to ask: How much money will I make? The answer to this question is always measured by the money you will receive—the returns—in one form or another, and we'll take a look at a few of the different ways these can be described in this chapter.

In order to compare deals and decide which ones to pursue, you need to know how the projected returns are calculated. When a sponsor tells you how much you're projected to make, there are two key things you need to know:

1. exactly what calculations the sponsor is using
2. the assumptions underlying the calculations.

Sponsors will use many of the terms that we are covering in this book. However, their understanding of how these concepts are defined may be different from yours or from that of other sponsors. While returns measure the relationship between how much you've invested in the deal and how much you may get back, the term might be used to describe the total overall profit for the deal or to explain how much money is being returned to investors at any particular moment in time. For example, they might be described the total nominal cash you will make on a project, in the form of a multiple of your investment, or project a snapshot of how returns will look at the end of the first year or some other specified moment in the project's lifecycle in the form of a percentage yield. Therefore, you need to get into the DNA of a deal to know exactly what they're projecting you'll be earning if you invest with them.

Ultimately, returns are a measure of profitability—a return on your investment. And while, obviously, everyone prefers higher returns, it is also important to remember that the higher the projected return, the higher the risk.

Returns measure the money that you'll make on your investment relative to the amount you've invested, and provided you use consistent definitions and calculations, you can use projected returns to compare one deal versus another, compare across asset types, or compare different geographical areas.

Defining and describing returns

There are four main concepts that help to define and describe returns:

- how much you'll make overall
- how much you'll make annually
- how much the entire deal makes
- the annual interest rate you'll earn

And there are four terms used in the industry to describe these different kinds of returns:

- return on investment (ROI)
- return on equity (ROE)
- rate of return (ROR)
- cash-on-cash

However, as mentioned above, there is no central school of real estate that every developer and sponsor goes to, which provides the *correct* definitions for these terms. Everyone understands these terms based on who or what they learned them from, along with their experiences.

If you have any doubts when you look at a deal about what the sponsor is referring to exactly, ask the sponsor to clarify what they mean when they talk about any of these terms—ROI, ROE, ROR, and cash-on-cash might mean something different to each sponsor. Don't be afraid to ask them directly what they mean or how they're calculating their numbers. Let's examine these different terms one by one.

Return on investment

ROI can be at the investor level, defining how much you'll make as a percentage of your initial investment, or it can describe the overall return for the deal as a percentage of total cost unrelated to total equity.

So, for example, if you invest $100,000 in a deal and you receive back a total of $150,000 from all sources (preferred return and profit split), then the ROI would be 50 percent. At the deal level, to calculate ROI you would take the total cost of the project, say $10 million, and total return to the project over the course of its life; assuming that was $12 million, the ROI at the deal level would be 20 percent.

Return on equity

ROE is a bit more specific, as it refers to the profits returned on invested capital in percentage terms—the total return on the equity portion of a deal.

While ROI and ROE are often used interchangeably, neither takes time into account. Instead, they are absolute numbers that can be used not only to define overall returns on a deal once it has completed its entire life cycle, but also to measure returns at a particular instance in time—a deal might be projected to yield 10 percent at the end of its life cycle or an 8 percent ROI at the end of Year 1.

Rate of return

In contrast to ROI and ROE, the ROR (while a similar calculation) delivers a measure of returns that is averaged over time. Using ROR, a sponsor will be attempting to explain what the returns looks like on an annual basis. For example, a 50 percent total ROI over five years might be described as being a 10 percent ROR per year.

Cash-on-cash

Cash-on-cash is how much an investment makes expressed as a percentage relative to the total amount invested at any given time. For example, if you invest $50,000 today, and the total trailing 12-month pretax cash flow paid to you at a given moment in the project's life cycle is $5,000, then your cash-on-cash return at that moment would be 10 percent.

To calculate the cash-on-cash return, the total cash investment is divided into the total gross rents less all expenses, including a deduction for financing costs (any amortized loan payments).

Speaking different languages while using the same terms

Because these four terms are used interchangeably by sponsors and can have different meanings, let's take a look at some specific examples to illustrate the underlying concepts.

DEAL 1—NO DEBT

Imagine we have a project with total cost of $20 million and a $1.4 million NOI in Year 1. The sponsor paid all cash for the deal and didn't borrow a penny from the bank. In this scenario, the ROI is 7 percent—$1.4 million divided by $20 million.

The equity is 100 percent in this debt-free deal, so the ROE is also 7 percent. As the NOI is an annual measure, and assuming a one-year deal life in this example, then the ROR is 7 percent. Finally, the cash-on-cash return is also 7 percent, because, with no debt service, cash-on-cash is calculated by dividing the $20 million investment into the $1.4 million of cash flow.

DEAL 2—SAME AS DEAL 1, BUT WITH DEBT

Using the same scenario, let's see what happens when we add debt, and assume *ceteris paribus*, which means other variables stay the same. Let's also assume (to keep the numbers easy for the purpose of this example) that the bank lends at 0 percent interest.[5]

Assume that the sponsor gets a $10 million bank loan, so we have the following:

- debt—$10 million
- equity—$10 million
- NOI—$1.4 million

This will yield the following returns:

- ROI—$1.4 million / $10 million invested = 14 percent
- deal-level ROI—$1.4 million / $20 million = 7 percent
- ROE—14 percent
- cash-on-cash (with no debt service)—14 percent

In this example, we now have a $20 million purchase and a $10 million loan. This makes the equity in this deal $10 million, and assuming the NOI of $1.4 million remains the same, the ROI is $1.4 million divided by $10 million, which yields a 14 percent ROI.

The unlevered ROI is sometimes referred to as the deal-level return; in this case, it remains at 7 percent. Similar to the cap rate, with deal-level return, you take out the influence of debt and all its possible variations in order to be able to compare one deal on more equal terms with another.

DEAL 3—WITH A SALE

Now let's add yet another layer to this scenario. Let's assume the building is sold. We buy it at a 7 cap for $20 million, and in this deal, the sponsor borrowed $10 million. We invested $10 million, or some proportion thereof, in equity. The NOI coming from this deal is $1.4 million. And the sponsor is able to increase revenues through Year 2 to $1.8 million. The market has stayed solid at a 7 cap, so the property is sold at that cap rate for $25.7 million, which gives us a capital gain of $5.7 million. In summary:

- debt—$10 million
- equity—$10 million
- NOI—$1.4 million
- increase in NOI in Year 2 to $1.8 million
- sale at a 7 cap—$25.7 million
- capital gain—$5.7 million

What is our deal-level ROI? To calculate this, three forms of income are taken into account:

- $1.4 million in NOI in Year 1
- $1.8 million for Year 2 NOI
- $5.7 million in profit when the building is sold

In this scenario, the total ROI is $8.9 million divided by (for deal-level profit) $20 million, giving us a 44.5 percent deal-level ROI. If that takes one or two years, it would be a very different return profile than if it were to take 20 years to get there. So while ROI can be a good indicator of whether a deal is worth pursuing, it does not paint the whole picture.

The deal-level ROI is calculated as follows:

- $1.4 million + $1.8 million + $5.7 million = $8.9 million
- $8.9 million / $20 million = 44.5 percent
- deal-level ROI is 44.5 percent
- returns are 144 percent or 1.44× multiple

For the investor ROI in this example, divide the $8.9 million by the $10 million investors have in the deal to get 89 percent ROI.

- $8.9 million / $10 million = 89 percent

While it's one of the most important things to know, sponsors may use ROI, ROE, cash-on-cash, ROR, or even 'yield' interchangeably. If you aren't sure and want to know what a sponsor means by a number, ask. You will probably get an interesting answer, maybe one that you didn't expect. At the least, the response will provide you with insights into the sponsor's way of thinking.

Financial key 4: Equity multiple

Equity multiple is more consistently defined across the industry and easier to calculate than other measures. The equity multiple is an important component for understanding how well a deal compares to other deals, especially when used alongside the IRR, which we will cover later.

Calculating the equity multiple

The equity multiple is calculated by dividing the total amount invested in a deal into the total cumulative returns, including ongoing distributions from rents, profit from sale, and return of investment. It is represented by the following formula:

$$Equity\,multiple = \frac{Total\,cumulative\,returns}{Total\,amount\,invested}$$

Basically, that is the total amount returned to investors divided by the total amount invested. It is an absolute number that doesn't take time into account.

Here's a breakdown of how total cumulative returns are calculated. Ongoing cash—whatever the cash was that came out of a project—plus the profit from the sale, plus the return of the initial investment:

$$Total\,cumulative\,return = Total\,income\,from\,cash\,distributions\,(preferred\,return)$$
$$+ \,Profit\,from\,sale + Return\,of\,initial\,investment$$

Let's say you invest $10,000 in a project that produces $500 in profit from rents, your share of the cash available for distribution from the project. The project sells and you get a $2,000 share of profit from the sale, plus the return of your original investment of $10,000:

$$\$500 + \$2,000 + \$10,000 = \$12,500$$

Assuming that you put $10,000 into the deal, that would mean the total gain the cumulative returns would be $12,500 over $10,000.

$$\frac{\$12,500}{\$10,000} = 1.25$$

In this scenario, your $10,000 would have an equity multiple of 1.25×. Stated another way, you get 1.25 times what you put into the deal. Is that a good return? If you make a 1.25× on your money in one month, risk excluded from the scenario, yes, it probably is a good deal. But what about over the course of a year, or even five years?

This is an important concept to understand, because it speaks to your investment strategy. Are you hoping to build wealth or earn passive income? If you're looking to build wealth, then you might want higher equity multiples over a longer period of time. If you're looking for passive income, then you should look at how the project delivers yield over time; in other words, what kind of consistent cash flow will come out of the project over five years, ten years, or even longer.

Financial key 5: Internal rate of return

The IRR is poorly understood, even in the professional investment community, and for good reason as it is a difficult concept to grasp. First of all, let's start with a dictionary definition of internal rate of return.

> *The discount rate at which the net present value of a set of cash flows, i.e. the initial investment expressed negatively, and the returns expressed positively, equals zero.*

In simpler terms, IRR means how much you will make over time, annualized, in percentage terms while taking into account the compounding effect of growth—that is, where the incremental growth of an investment from one year grows the following year.

To stress how complicated IRR is, I have worked with hundreds, if not thousands, of professional real estate developers, sponsors, and investors, and a lot of them really don't know what it means. In many cases, they don't really care. In fact, many investments happen without anyone knowing what the IRR is, why it matters, or why it should matter. Nevertheless, it is a key concept and one that is commonly used in commercial real estate and in real estate crowdfunding.

The IRR, though not without its limitations, serves as a useful comparison tool, as it can be used to understand how a deal will perform and how its risk-return profile compares with others.

The IRR's function is to calculate returns on an investment for multiple periods. When you examine all the revenue streams and all the different moments during a project's life cycle when money was invested in the project, the IRR will tell you the overall return, on an annual basis, based on all revenue streams including cash flow and proceeds from sale.

Who uses IRR?

The IRR is among the most frequently used measurements of overall returns, but it still remains completely unused in many circles. In my own experience, I've dealt with sponsors who simply do not use the IRR. They only look at equity multiple. For them, they don't care if a deal takes two, three, four, five, or six years as long it doubles their money. They focus on the overall return as a multiple of how much they put into a deal and then what kind of cash flow a property will produce.

Long-term, buy and hold investors also tend not to be too focused on IRR, because they are thinking so far into the future that making realistic projections are unrealistic and they never anticipate a terminal value (a sale), which is necessary to calculate the IRR. Their investment strategy—to patiently wait for assets to build wealth over the long term while adding to income streams and using low debt—doesn't need an annualized projection of returns. These investments rely more heavily on core fundamentals and a desire to hold on to a property indefinitely.

In other circles, particularly in the institutional world, IRR is a mandatory calculation. It is commonly used by sponsors crowdfunding their syndications. You must be able to understand what the IRRs are, because it's an excellent metric for either understanding your own risk-return profile or the risk-return profile of a particular deal; and yet it is also subject to distortion.

IRR and target return

IRR links future returns to the initial equity investment. It reflects future returns relative to the investment you make today—the actual equity you put into a deal; in other words, how much you will make on an annualized basis, going forward, in an accurate, time-adjusted way.

IRR is also used for calculating capital allocation in deals and initial investment decisions. For instance, you might have a target annual return that you want on your investment—you want to make a minimum of, say, 10 or 15 percent per year; the IRR helps to project forwards what the return will be for any particular deal so you can compare that deal against your own target returns.

The IRR is also useful because it varies according to the degree of risk and factors such as real estate type and market cycle. Below is an example which illustrates this concept.

Investing in a Class A office building, one of the lowest-risk investments in real estate, might carry a single-digit IRR. If a project has some value add potential, like in a core plus deal, it might offer a high single-digit return. For heavier value add deals, it could possibly carry an IRR in the mid to high teens. And with ground up deals, the projected IRR can be upwards of 20 percent or more. In other words, the IRRs are positively related to the amount of risk you have in a deal—the more risk, the higher the projected internal rates of return you would expect to compensate for taking on that additional risk.

Remember: Based on the four investment strategies we covered earlier in this book, the more work needed to stabilize a project, the higher the risk that something could go wrong; and as risk goes up, so does (so *should*) IRR.

The IRR paradox in crowdfunded syndicates

However, the IRR range seen above—single-digit for Class A office building all the way out to 20 percent plus for ground up development—will change depending on the stage of the real estate cycle. In the very early stages of the uptick of a cycle, right after a recession, the projected IRR across the risk spectrum will be higher than near the end of an up-cycle. Why? Because it becomes increasingly difficult to find deals with good returns as the market cycle progresses.

But there is a paradox that has emerged with the advent of crowdfunding where IRRs are going up when they should be going down. This is how it's happening. After a downturn, when investors need elevated incentives to return to the market with their capital, because losses are still fresh in investors' collective memories, sponsors must find deals that offer higher returns to compensate for what investors see as heightened post-recession risk. Finding these kinds of higher-return deals during post-recession times is easier than it is during a healthy economy, because there is a greater prevalence of distressed deals for which there are more ways to add value to a project and where assets are often undervalued.

As the real estate market improves and distressed projects become less prevalent, it is inevitable that the kinds of opportunistic returns found earlier in the recovery will become increasingly difficult to find or increasingly expensive to buy, compressing returns. This, in turn, leads to a gradual reduction in projected IRRs as the market returns to its natural state with increasing supplies of liquidity (from both lenders and investors) and competition heating up, which drives down pricing and returns.

The **crowdfunding paradox** is that as market forces are driving IRRs downward, crowdfunding is driving them upward.

Any seasoned investor understands that as a market recovery gets longer, IRRs will go down. But this happens in reverse in crowdfunding, because there are too many unsophisticated investors who demand ever increasing IRRs in the belief that the market will always go up, and there is always a supply of sponsors willing to feed that greed.

Remember the story of the developer who would build an apartment building in the middle of the desert, where no one wants to live, just because investors are willing to give him the money to build it?

This is the same thing, another case of the tail wagging the dog. In crowdfunding, the higher the projected IRR a sponsor shows, the more investors congregate to learn more. They demonstrate this by showing up at webinars in greater numbers for those sponsors indicating the highest IRRs, and by clicking on those projects with greater frequency and spending more time examining the details.

Think about it; it's a natural human instinct, a little lemming-like admittedly, but one that feeds our natural instinct to beat out the competition. For the first time in history, investors can see private real estate deals offered to them on the same page, alongside each other, like groceries on a shelf. And with groceries on a shelf, when all the products look alike, the first thing most prudent shoppers do is look for the least expensive option on a price per pound basis.

I do this all the time when shopping for that basic staple, M&Ms. These come in the small 6 oz snack pack, the 1 lb family pack, and the mega 2 lb party pack. The only way to make a decision is to look at the relative price per ounce—which isn't always the mega party pack, I'm afraid to say. Why would you pay more for less? If buying two 1 lb bags costs less on a price per pound basis than buying one 2 lb bag, buying two 1 lb bags is the only logical option.

The same hunt for values drives the crowdfunding investor. Lined up on a digital shelf, as they are online, looking for those deals that offer the best return for the investment makes all the sense in the world—except it can also be hazardous to your health, just like buying those two bags of low-cost, sugar-packed, calorie-stuffed, chemical-laden candies and taking them home to gorge on them.

In the competitive environment of an online shelf of projects, by demanding higher projected IRRs from developers, investors are acting counterintuitively to the natural flow of the market—and yet they are sending so powerful a signal that neither the platforms nor their sponsors can resist.

The nature of online crowdfunding is that it informs sponsors that they must offer higher IRRs or their projects will not be financed. Sponsors know that they must have higher IRRs to compete shoulder to shoulder online.

This encourages sponsors to become more liberal with their assessment of projects, willing to buy at higher prices, project lower costs and higher revenues than might be realistic, and squeeze the project life cycle into unrealistic periods, all in the name of cranking up projected IRR to compete more effectively with other sponsors all doing the same thing.[6]

Impact on the promote

There's another way that the IRR can distort the success of the deal. When the promote is tied directly to the IRR, as it is in 90–95 percent of cases, it creates pressure on sponsors to exit a project while leaving money on the table, because the effects of selling a project earlier than projected is that there is an increase in the IRR. If a sponsor has a promote with a 20 percent hurdle, they can achieve that hurdle in one of two ways: by pushing down cost and/or increasing rents and squeezing value out of a project; or by selling it ahead of pro forma exit dates.

I saw similar situations while working in private equity. During the financial crisis of 2007–2008, the Federal Deposit Insurance Corporation (FDIC) was selling vast portfolios of nonperforming mortgages from banks that had failed. The buyers were typically private equity funds that were purchasing these real estate collateralized loan portfolios based on projected values and exit dates for when they would sell the loans at a profit. When they underwrote the acquisitions, they solved for target IRRs. Executive bonuses were dependent on achieving those IRRs and were increased if they were able to exceed the targets. Simply by finding buyers for these portfolios earlier than was originally projected, it was possible to spike the terminal IRR, and bonuses were larger than anticipated and were paid earlier. Everyone did well, except that money was being left on the table. Once it became apparent that there was more value to be extracted by holding on rather than selling early, these kinds of adventures were shut down. There was nothing nefarious about how this worked. The rules were agreed by everyone in advance: provided certain IRR hurdles were met, assets could be sold. It wasn't until after the first few portfolio sales had occurred that the exit strategy was reassessed and a new approach was adopted.

Sponsors in a private real estate deal are faced with a similar quandary. They may know that rents could go higher or that costs could go lower. They may really want to hold on to a project for the long run, but being motivated by the IRR for the bulk of their own compensation, they may be tempted to sell earlier in the life cycle of a deal than would benefit shareholders overall.

Again, this isn't necessarily driven by some kind of nefarious plot to deprive investors of returns. If you invest in a deal that exits earlier than expected, receive a check along with a statement that you just earned 20 percent IRR instead of the 15 percent IRR you were expecting, in all likelihood you would be a happy camper. It's another case of the tail wagging the dog, only in this case everyone benefits—though not necessarily as much as they could have done had they stayed the course and if they had used a metric other than IRR, or in conjunction with it, to measure the promote-based hurdles.

Calculating the IRR

The IRR is not perfect and, as with other ways to measure performance in a real estate deal, should be used alongside the other eight financial keys to paint an overall picture of how a project stacks up against others and whether it is right for you.

That said, now that I've gone into so much detail on how the IRR works and how it is flawed, it's probably a good idea to explain how it is actually calculated. In short, the IRR measures the overall return on your investment over time, measured annually, in a way that takes compounding into account. Compounding, as I mentioned earlier, involves returns from one year themselves seeing a return the following year, and so on. Put another way, think of interest you earn on interest in a bank. If you leave your bank deposits in place, then each year the interest that accumulates in your bank balance itself earns interest. If you keep that idea in mind, you will have captured the basic mechanism behind which the IRR is calculated. Let's unwrap that now by looking at some examples.

Scenario 1

Let's say you make a $1,000 investment in a deal, and in five years you get back $1,500. Your equity multiple in this case—your total profit—is 50 percent or 1.5× your invest-ment. Is this the same as 10 percent per year return on your money? Would it be a straight calculation? Fifty percent over five years: 50 divided by 5 equals 10 percent return per year, right? Not really. This calculation does not accurately reflect the actual annual return on your investment because of the way interest compounds over time.

YEAR 1

With $1,000 invested in the deal, you earn $100 (10 percent).

$$\$1,000 + \$100(10\%) = \$1,100$$

YEAR 2

You earn ANOTHER $100 on the $1,000 PLUS 10 percent on the $100 earned in Year 1 ($10), for a total of $110. You now have $1,210 in Year 2.

$$\$1,100 + \$110(10\%) = \$1,210$$

YEAR 3

You earn 10 percent on the $1,210 you already have in the deal, adding another $121, which gives you $1,331.

$$\$1,210 + \$121(10\%) = \$1,331$$

The good news is that this keeps going. In other words, the 10 percent that you earn each year also earns 10 percent for you, and this compounds over time. This is why the initial example is not a simple 50 percent over five years to give 10 percent. The actual IRR in this scenario of a $1,000 investment with you receiving $1,500 at the end of five years produces an equity multiple of 1.5—a 50 percent profit—and in this case, the IRR is 8.45 percent. If you had invested that $1,000 in a bank account that paid 8.45 percent, at the end of five years, you would have received exactly $1,500.

Scenario 2

In this scenario, you invest $1,000 to get into the deal. In Year 1, there's an option to invest another $500. In Year 2, there's an option to invest another $600 due to expansion or other opportunities occurring around the deal. You invest:

- $1,000 to enter the deal
- $500 in Year 1
- $600 in Year 2
- nothing in Year 3

In Year 1, your income is $100, but there is nothing in Year 2. However, since you're investing, you earn $150 in Year 3, then $175 in Year 4; and in Year 5, the project sells for $2,000. What's your return in that scenario? That's where calculating the IRR on your investment comes in, because it allows you to calculate a rate of return on a project even when there are multiple points of investment over time and varying levels of returns in the same period.

Going back to the earlier investment scenario, imagine an initial $1,000 investment, an additional $500 invested in Year 1, $600 in Year 2, and zero invested in Year 3. Using the income projection above, the IRR in this deal is 3.83 percent. Is that actually something you'd want to do? The decision depends on what your goals are and how much risk is involved. And, specifically, it depends on your risk-return analysis and your appetite for risk.

Plus, if you have specific targets for how much you want to earn on your investment, the IRR can be used to determine whether a deal is going to meet those targets. And it helps you figure out if the deal is competitive by helping you to compare it with other deals.

In short, the IRR provides a useful method of measuring overall returns and comparing deals against each other even if those deals have complicated structures, varying investment and payout schedules.

Time value of money

Underpinning the calculation of the IRR is the concept of the time value of money, which is measured using a discounted cash flow analysis. But don't worry. The time value of money and the idea of a discounted cash flow is not something real estate pros actually talk about. Nobody says, "I like that deal because it has a great discounted cash flow." I've never heard anyone ask, "What is the discounted cash flow in that deal?"

In fact, truth be told, I've only ever heard this term used in academia and textbooks, where it is discussed to the nth degree along with totally impenetrable calculations that are pored over to derive the IRR. Here in the world of mortal real estate professionals, we are blessed with the wonders of modern science—aka Microsoft Excel—with which it is simple to calculate the IRR of a deal without needing to have an inkling about what it is or how the calculation works.

But, of course, therein lies the problem also. As the evangelist Billy Graham is reputed to have said, "technology merely gives us a more efficient means of going backwards." Virtually no one has a clue what the IRR really is in the context of the time value of money (ask any real estate pro if you don't believe me). So let's take a look at this important concept. It will give you a foundational understanding of how the IRR works as well as providing you with an interesting topic to discuss at cocktail parties.

To explain time value of money, think about which of the following you would prefer. Option 1: I could either give you $100 today or $100 in a year's time? Presumably you would take $100 today because you could invest that $100 in a real estate deal and maybe turn it into $110 in a year. That $100 is worth more to you today than a $100 would be in a year's time.

What if I gave you Option 2—$100 today or $110 in a year's time? Which would you take? You might take the $100 today, or you might say you'll wait for the $110. If you take the $100, it may become $110 in a year. But that might not be tempting enough. Maybe you've found a deal on a crowdfunding website that is offering an 18 percent yield. You might think, "Well, I'm not going to wait for that $110. I would rather take the $100 now, invest it, and earn $118 in the online deal."

Taking that to the next level, how do you make that same decision if there are multiple years involved? In this scenario, assume you have a target return of no less than 12 percent on your investment. In the example above of $100 now versus $110 in the future, would you take $112 instead? But what if the numbers looked like this: There's a prospective deal that involves investing $1,000 today. In Year 2, you get nothing. In Year 3, you earn $200. In Year 4, you earn $300, and so on. Using an IRR calculation, you can determine whether the overall return you'll receive is compatible with your target of a 12 percent net return on your investment.

In the same way that the IRR can be used to provide guidance to an investor to determine if a deal meets certain return hurdles, so it can be used by sponsors to solve for a return target they want in order to determine if a deal is worth pursuing. Instead of using total predicted investment and returns over the life cycle of a project to calculate the IRR, determining how to 'solve for' an IRR requires a slightly different approach.

Solving for a certain IRR means adjusting the variables in a deal to ensure a certain IRR is met. For example, a sponsor will underwrite a deal based on predicted costs and revenues, leaving the purchase price of the asset as a key variable. If, alongside what the sponsor believes to be realistic costs and revenues, the asking price yields an IRR that is unacceptable to them, they can adjust the remaining variable, purchase price, until the IRR they want is returned in their calculations. Let's say, arbitrarily, that an asset is offered at $10 million and, after the sponsor has underwritten projected expenses and income, the project pencils at a 15 percent IRR; but the sponsor wants a 20 percent IRR, so they will vary the price variable until their projected IRR equals 20 percent and use that number to make their offer. This is an example of how a sponsor would 'solve for' a target IRR.

Think of it this way: If you're strictly looking at the time value of money of your investment, then the question you want to answer is, for example, how much is $100 in three years' time worth today? This is similar to asking how much could a dollar today have bought yesterday?

To illustrate these differences, let's look back at Table 2.1 in Chapter 2, which compares 2017 prices and 1980 prices. I chose 1980 as the base year to illustrate this point because that's when the accredited investor standard was established (promulgated in 1982), so this comparison also serves to illustrate how while the standard has remained largely untouched since adoption, economic times have changed dramatically. This speaks to the enormity of the opportunity crowdfunding real estate syndications presents to both investors and developers.

In that table, we can look at several items to compare how the time value of money can differ depending on what variable you're looking at. How much a dollar in 1980 would be worth today depends on the discount rate, which is the same as the IRR, and the discount rate depends on a number of factors.

When you look at the Dow Jones index, $1,000 in 1980 would be worth $22,118 today. This increase is by a multiple of 22. But then you have to take the inflation rate into account; this was 13.5 percent in 1980, and, as of 2017, 1.7 percent. If the inflation rate rises at the same rate as the Dow Jones, all you're doing is sitting still. You're not improving your wealth or earning any kind of real income.

By the same token, if you'd bought a home for $76,400 in 1980—the national average—that same home would be worth $200,000 or more today. The discount rate, compared to the Dow, is much lower. Homes went up by less than 3 times in the same period as the Dow increased by 22 times.

How do you calculate the value of a dollar today as it would have been in 1980 when there are such huge variances in the pace at which it would have grown, depending on the benchmark being used for comparison or, to put it another way, depending on the discount rate being used? Conversely, how do you calculate the discount rate on a future stream of income based on the dollar you invest today? If a sponsor claims that they're going to return you a certain amount of money in five years, how much is that stream of income, and return, at the end of five years really worth to you?

How do you figure that out? What discount rate will you use? Look at several items in the table, like postage stamps, or gas, or minimum wage. If we focus on minimum wage, we see it has actually gone down in real terms.

Using baselines

To resolve this quandary, a common practice is to use economic baselines, which represent zero-risk investment returns. Since the lowest-risk investment you could possibly make is in US Treasury bonds, the rate at which you can buy these instruments is used as a baseline for calculating a discount rate, upon which the time value of money is founded. Think of it this way, if you could invest in something at zero risk, like US Treasuries bonds, and earn 2 percent, there would be no reason to invest in something else that carried risk only to earn that same rate. As real estate cycles lengthen and prices get higher, returns on investments get smaller; and as that happens, the gap between what you can get in a real estate investment and with zero-risk Treasury bonds narrows. The relationship between returns in real estate (or any other investment) and zero-risk Treasury rates is a measure of an individual investor's risk-return profile; the bigger the difference, the more tolerant an investor is to risk.

An example: A land purchase

Let's look at an example that puts the time value of money concept to use by examining how to calculate the value of a piece of land in order to make an offer on that land.

Imagine a sponsor finds a land development deal that's on the market for $3 million and is looking for a 20 percent IRR return on their investment. Doing a 'back of the envelope' analysis, he figures the construction costs for what he wants to build will be around $17 million. Through a feasibility study, he also determines that his stabilized NOI will be $1.4 million once the project is completed. In addition to that, he projects the exit cap to be 5 percent, or an exit value of $28.4 million.

That 20 percent return that the sponsor wants covers the deal's risk premium, which includes default risk, market risk, construction risk, liquidity risk, and credit risk, plus, of course, profit. The sponsor builds all these risks into his projected IRR on the money he will invest today. So, if he's solving to a 20 percent ROI, the sponsor will vary the price of the land, taking into account all the underwritten and predicted income, expenses, and investment capital, until it returns a 20 percent IRR. And that number will be what the sponsor will be able to afford to pay for the land and will underpin his offer to the seller.

Now that we've considered how sponsors use the concept to discount rates, let's go back to IRR. Just remember that it's a measure used by some—not everybody—to create an overall view of the deliverable returns in a deal.[7] As explained above, the IRR is used to compare one deal against another, allowing for an apples-to-apples comparison of all deals and all types of real estate investment. The IRR gives a consistent perspective of the overall returns on a deal by looking at the annualized returns, but it's different from cash-on-cash returns. The IRR really explains how hard your money is working for you across a range of very complicated circumstances. This is what makes IRR consistently useful across every single type of real estate investment you might consider.

The IRR variance ratio

At the deal level, the IRR is going to be different than the investor-level IRR due to deductions for sponsor fees and promotes that occur before you get paid. The size of this difference is also illustrative of whether or not this will be a good deal for you, the investor.

The 'IRR variance ratio' is a powerful tool for you to use in comparing one deal with another, because the difference between the deal IRR and the investor IRR can only be explained by the money that the sponsor is taking out of the deal for themselves. This includes the money they are taking out in their incentive payments, like preferred return and promote, but does not include fees, because those also are deducted as an expense to the deal-level IRR. By finding the difference between the deal-level IRR and the investor-level IRR and dividing that by the investor-level IRR, you get the IRR variance ratio:

$$\frac{Deal\text{-}level\ IRR - Investor\text{-}level\ IRR}{Investor\text{-}level\ IRR} = IRR\ variance\ ratio$$

The lower the IRR variance ratio, the less the sponsor is taking out of the deal (other than in fees) compared to the investors (which is good, obviously); the higher the ratio, the more the sponsor is taking out (not good).[8] As a standalone ratio on any single deal, this doesn't mean very much. But when you use this variable to compare multiple deals, you can see how one deal stacks up against another. I have conducted an extensive comparative analysis across dozens of deals and platforms and found that this ratio can vary from the high single digits to the low 30s. As it's an easy ratio to calculate, it's a useful tool you can use.

Timing a factor

When comparing different deals, remember the timing of payments is key to how the IRR is measured and calculated. Varying the exit period in IRR calculations can dramatically alter how the IRR will look. For example, a project that projects a 15 percent IRR over five years may show a 20 percent IRR if projected over only four years. So in a world where investors are driven by IRR, it can be tempting to underwrite projects to exit faster than they might otherwise in order to spike the headline IRR to draw in more interest—as discussed earlier.

In conclusion, the IRR is probably one of the most useful tools that you have when looking at a deal. Use it to compare one deal against another and base it against the kinds of returns you might get. Or compare it against the kind of target returns you might have for your investment dollars.

Financial key 6: The waterfall

In every deal, money, in the form of debt and equity, flows in and out of the project's finances over time. The way in which money flows out of a deal back to its lenders and investors is a hierarchical process that is determined by what is called the waterfall.

If you own a home, you are probably already familiar with the simplest of waterfalls. When you purchased your home, you likely borrowed from a bank by applying for and receiving a mortgage as well as making a down payment, because the bank wouldn't lend you 100 percent of the purchase price of the home. When you sell the house, the first payment made from the sale of the property goes straight towards paying the bank back everything you owe, including the bank's interest, and any remaining principal borrowed (the base amount of the loan). After the bank gets paid, you get your down payment back plus any profits if the house went up in value. You also suffer the losses if the house goes down in value, by receiving back less than the down payment you made. If the value of the house went down more than your down payment, the bank still expects to be paid, and you must come out of pocket to pay them back or suffer considerable financial consequences through foreclosure or bankruptcy and the inevitable impact either of these scenarios will have on your ability to borrow again in the future.

This order of payment—first to the bank, next to you—describes the waterfall structure of your home purchase. If you owe money to contractors who did work for you on your home, they also form part of the waterfall of homeownership; they will take priority over you but come after the bank.

I don't know how the word 'waterfall' came to be used in this context. The word itself conjures an image of water cascading in a downward direction, and I suspect it is related to the way calculations are laid out in a spreadsheet. At the top of a spreadsheet is income, below that expenses, followed by NOI. Lower down the spreadsheet are all the distributions. Below the NOI line, debt is taken out, and below that, the next line of the spreadsheet is assigned to whoever has the next priority for payment. In the case of your home sale, that might be contractors or other lien holders you owe money to, and then after them, next down on the spreadsheet would be money to which you are entitled.

Here's how it looks in order of priority when you sell your home.

1. You get income from the sale.
2. You pay expenses related to the sale (brokers, title companies, etc.).
3. The bank gets paid whatever interest is owed.
4. The bank receives the principal balance owed to it.
5. Lien holders get paid.
6. You get your down payment back.
7. Finally, you get whatever profit (or loss) there may be for selling the house for more (or less) than you paid for it.

When you invest in a real estate deal online, you'll also become part of a waterfall, but one that is likely to be a lot more complicated than the homeownership waterfall. We'll take a look at those now.

There is one important similarity between the homeowner waterfall described above and any investment waterfall, and one way in which they become considerably more complicated. The commonality across all waterfalls—with only some very rare, technical exceptions—is that the lender always sits at the top of the waterfall and gets paid first.[9] Where they can get complicated is when investors are involved in a project. Imagine, if you will, that in the simple homeownership example above, instead of being the only person to purchase the home, you joined with partners to buy it, perhaps to operate a vacation rental together, each contributing a share of the down payment, signing on the loan (guaranteeing the loan), and sharing in the operational responsibilities of running the rental business. You would have to agree specific terms on how money is distributed out of the project, both on an ongoing basis from the profits of operating the rental business and for when the home is sold.

Taking this one step further, consider a scenario where you wanted to buy and operate a vacation rental yourself but did not want partners to operate the business with you, only investors to help you with the down payment. In deciding how to compensate yourself and your investor partners, you would be designing an increasingly complicated waterfall structure—a commercial real estate investment waterfall.

In the commercial investment world, the simplest waterfall I've seen is what I call the 'friends-and-family waterfall.' In this structure, the investors contribute all the money that a deal requires. The sponsor puts in nothing but, in these friends-and-family deals, neither do they take out any fees. When the deal sells, the sponsor and investors split the profits 50/50. It's a clean, simply structured waterfall. Remember that in any waterfall, the bank is always the first party to get paid back, so this friends-and-family structure looks like this.

1.   The bank lends debt.
2.   Investors contribute all the equity.
3.   The sponsor contributes just 'sweat equity' and none of their own capital.
4.   The sponsor takes no fees.
5.   The building sells.
6.   The bank gets paid back.
7.   Investors receive back all their equity.
8.   The sponsor and investors split the profits 50/50.

These kinds of no-fee, straight-split structures are rare. I've only ever seen them used by family offices and other especially close-knit investor groups. The vast majority of waterfalls include a 'preferred return' that is paid to investors before sponsors receive any share of the profits, and after the bank (always). To balance that, however, sponsors, almost without exception, take fees of some sort to cover their overhead costs of operating a real estate business.[10]

The preferred return, often referred to as the 'pref,' is similar in some ways to the interest you get on a deposit when you put your money into a bank—with the important proviso that your money deposited in a bank account and the interest earned on it is generally guaranteed by the federal government, whereas the money you put into a real estate deal and the associated preferred returns are not guaranteed by anyone, and you may lose all your investment.[11]

When you're looking at how much pref you'll receive, consider the developer's fees before you start earning money on your interest to ensure there is an alignment of interest between you and the sponsor. We'll return to the preferred return and the concept of the alignment of interest later.

Waterfall versus capital stack[12]

The flow of money through a waterfall is in the exact opposite direction of how it flows in a capital stack. It's like a crafty real estate industry ploy to deliberately confuse everybody involved. Here's how they are different. In the waterfall, the highest-risk, highest-return capital—your investment—is situated at the very bottom of the flow of capital, whereas in a capital stack, your capital is represented at the very top of the stack. The waterfall and the capital stack are essentially describing the same thing; the hierarchical way money flows through a project and the relative priorities the contributors of the various types of capital have over each other. But they do it in very different ways.

Understanding alignment of interest (and why you should care about it)

The concept of alignment of interest is incredibly important. It is designed to ensure that both your interests and the developer's interests are compatible, that they are aligned. The developer must be motivated to make money in a manner consistent with serving your best interests. That is the definition of the alignment of interest.

Let's look at an illustrative scenario. In a $1 million deal, the bank lends $700,000 while the investor puts up all the $300,000 of equity. In this example, let's allow the developer to take fees of $75,000 for putting the deal together. The investors are paid back their investment when the deal is sold, and profits are split. This is how the deal looks.

1. The bank lends $700,000.
2. Investors put up all of the necessary $300,000 of equity.
3. The developer takes fees of $75,000 for putting the deal together.
4. Investors are paid back their investment when the deal is sold.
5. The profits are split.

In this example, the interests of the developer and the investors are not aligned, because the developer has absolutely nothing to risk in this deal *and* nothing at all to lose; they only stand to gain. In this scenario, the developer stands only to gain from fees and from the profit split, because they take 25 percent of the investment as fees before investors get their money back ($75,000 of $300,000); in contrast, only investors and the bank stand to lose money.

Introducing the preferred return

In the scenario described above, adding a preferred return might help to realign interests. Here's how that might look.

As in the example above, the bank still lends $700,000 and the investor (or investors) put in $300,000. Although the developer still takes their fees up front, before splitting the profits, the investors get a preferred return, say 8 percent, which is the most commonly used preferred return in the industry.[13] The pref is paid to the investors before profits are split.

Now that the preferred return is introduced, you can see the needle moving slightly more favorably towards the investors in this scenario. If this example deal were to last three years, investors would earn $24,000 per year in preferred return distributions for three years, totaling $72,000 over the course of the project's life cycle. Below is a summary of how this $1 million deal looks with the introduction of preferred returns.

1.   The bank lends $700,000.
2.   Investors put up $300,000 of equity.
3.   The developer takes fees of $75,000 for putting the deal together.
4.   Investors are paid $72,000 of preferred return.
5.   Investors are paid back their investment when the deal is sold.
6.   The profits are split.

In this scenario, the preferred return paid to investors almost equals the fees paid to the developer, so some sort of balance has been created between the two parties' interests. That said, importantly, the developer in this scenario is still paid fees up front out of the investor's capital and still stands to lose nothing. But you can see how payment of a preferred return begins to balance the alignment of interests, because investors are paid a pref before the developer receives any share of the profits.

A $30 million institutional-sized deal

In a typical institutional structure, there is usually a much closer alignment of interests than in the friends-and-family scenarios described above. Let's say we have a $10 million deal. In this scenario, the bank lends the deal $7 million. On top of that, there's a required investment of $3 million, to which the developer (the sponsor) also contributes a portion.

In institutional-caliber transactions, the developer will often invest anywhere from 2 or 3 percent to 10 percent of the equity in the deal, depending on the deal size, and they will bring in outside investors to make up their share of the capital requirement.

In this example, a sponsor will find investors who may put up 90 percent of the $3 million equity required to finance the project, leaving the sponsor needing to come up with 10 percent of the $3 million, so $300,000 of their own money. The sponsor's portion is usually treated the same as the outside investor's money. The term that's used for that is *pari passu*, meaning 'on equal terms' with the investors.

Incidentally, this is going to be a very common structure in crowdfunded syndications you see online. The sponsor will have raised debt from a bank and will need equity. They will come to you for that equity and will contribute some of their own money to the deal too.[14]

Here's how a typical institutional structure with close alignment of interests in this $10 million deal might look.

1.   The bank lends $7 million.
2.   The investment required is $3 million.
3.   The sponsor contributes 10 percent of equity, or $300,000.
4.   The investors invest 90 percent, or $2.7 million.
5.   Invested equity is paid a preferred return.
6.   The developer is paid a 'promote' plus market rate fees.[15]

The sponsor's investment contribution to the deal in this case is 10 percent of the total equity. This is variously referred to as the 'co-investment,' 'co-invest,' or as having 'skin in the game.' As an investor, you want the sponsor to have some of his own money in the deal because you want to see them risking some of their own money alongside yours on exactly equal terms.

In the waterfall structure illustrated above, there is a much closer alignment of interests than in the friends-and-family structure. The order of repayment in this scenario is as follows.

1. First, the bank receives back the interest owed.
2. Next, the bank receives back the principal of $7 million.
3. All investors, including the sponsor (usually), are paid any preferred return owed in direct proportion to their investment.
4. All investors (including the sponsor for their co-invest in their role as investor) are paid back their principal, the original investment.
5. The sponsor is then paid a 'promote' which is a pre-agreed percentage of the profits.
6. The remaining profits are split among investors in proportion to how much each party invested (pro rata). This includes the sponsor, according to whatever proportion they put into the deal—in this case, a 10 percent co-invest.

As you study actual deals in crowdfunded syndications, be alert for different ways that preferred returns can actually be structured and paid. For example, you might see deals offering what is called a 'preferred equity' position.

Preferred equity

Throughout the course of my career, I have seen preferred equity defined in any number of ways. Historically, and on the deals I've worked on, the term was used to describe a structure that in some way prioritized one equity position over others, where one class of investor received preferential treatment for their equity over another. We'll take a look at a few of those kinds of scenarios momentarily, but the term 'preferred equity' has also come to mean something quite different since the financial crisis of 2007–2008.

During the financial crisis of 2007–2008, banks uncovered unexpected ways in which their priority positions had been compromised that made it harder for them to foreclose on properties to preserve their first-position claims. One of the causes of these problems stemmed from additional layers of debt that had been placed behind the banks, and one in particular that was known as 'mezzanine' debt—the last layer of debt between the debt and equity positions.

Lenders who had extended mezzanine debt to projects were able to leverage influence over a project that was disproportional to their lien position or amount of debt. As banks realized that mezzanine debt could create these kinds of problems, they increasingly prohibited borrowers from taking on second-position debt behind them—making it much harder for mezzanine debt lenders to operate as they had done previously.

Enter preferred equity in a new guise to fill the void.

Like mezzanine debt, preferred equity is also subordinate to first-position bank debt; but, importantly, it is ahead of 'common' equity in the capital stack. Despite its name, preferred equity acts more like debt than equity, though with some wrinkles that keep it looking like equity to conform to the new banking restrictions on placing debt behind it.

Like other forms of debt, preferred equity is superior to common equity, but unlike common equity it does not share in the profits of the deal, receiving only an interest rate payment on its principal, which is paid before common equity gets its preferred return or share of profits.

What differentiates the new preferred equity from the old, apart from not sharing in profits, are the rights that belong to the holder of this form of equity and the way in which they can secure their position. Most importantly, preferred equity holders retain rights to take over a project in the event a borrower defaults on their payments—you will almost never see this in common equity crowdfunded syndications, although institutional equity investors will retain similar rights.

The difference between preferred equity holders and banks—even though both receive only interest payments for their loans with no share of profits—is that in the event of a default, banks will foreclose on the property, taking ownership of the underlying real estate, whereas preferred equity lenders take control of the operating entity, which is a much easier process than foreclosing on the real estate, as the bank would need to do.[16]

There are many variations of preferred equity and, indeed, of how preferred return can be paid to different classes of investor. You must be vigilant in understanding a deal's structure—who gets paid, what they get paid, when they get paid, and how they get paid—in order to ensure an alignment of interest between you and the sponsor and to know your rights in the event something goes wrong with the project. If a deal doesn't feel right or you don't understand how it's structured, move on.

Sample deal structures

Below are some deal structures that you might come across that describe the nuances in how equity can vary from one class of investor to another. These are preferred equity positions of the 'old' kind, which behave like equity rather than debt, as in the example above.

Sponsor equity is subordinate to investors

In this structure (shown in Figure 8.2), the sponsor contributes 10 percent of the equity and the investors contribute 90 percent, but the investors' capital is treated with a higher priority than the sponsor's. Profits are distributed to investors until they reach a relatively high 12 percent preferred return, and then they receive all their capital back. This is before the sponsor receives a single penny of profit share. Then there's a straight split, 20 percent to the investors and 80 percent to the sponsor.

The sponsor's got skin in the game, but isn't treated in the same way as the investor. Indeed, in this scenario, the sponsor receives no preferred return on their co-invest, nor do they

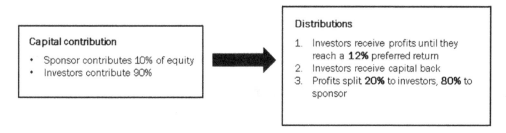

Figure 8.2 Sponsor equity is subordinate to investor equity

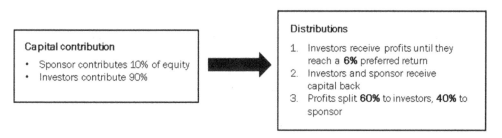

Figure 8.3 High pref, low returns, low risk for investors

receive any of their capital back until investors are made whole on their preferred return and original equity contributions; but the sponsor does receive 80 percent of the profits.

Higher preferred return to investors, higher share of profits to sponsor, lower returns and lower risk to investors

In this structure, the investors are paid back before the sponsor receives any preferred return or profit and before they get a return of their co-invest money. The sponsor is highly motivated to perform, because their share of the profits is so high and their own co-invest capital is at higher risk as it is not returned to them until investors are paid back. This is a lower-risk, higher-preferred-return scenario for investors, but with a lower expected payout because the profit split strongly favors the sponsor.

This same structure might be tweaked slightly (Figure 8.3), where the sponsor still contributes 10 percent of the equity while the investor contributes 90 percent, but the preferred return of 6 percent is lower than in the prior example and the 60/40 profit split favors the investors over the sponsor. In this case, the sponsor receives no pref but does get their capital contribution back alongside the other investors, after the other investors have been paid their pref.

Lower preferred return to investors but higher share of profits, higher returns, and higher risk to investors

Here, the investor receives profits from the deal until they receive a 6 percent preferred return. The sponsor receives no preferred return. Then, the investors and the sponsors

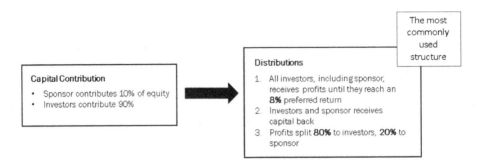

Figure 8.4 Most common waterfall structure

receive their capital back. The profit split is 60 percent to the investors and 40 percent to the sponsor.

This structure presents slightly higher risk but higher return to the investor. There's a lower preferred return than in the first scenario but a bigger split of the profits—plus the sponsor's equity is treated more in alignment with that of investors.

Most common waterfall structure

Historically, the most commonly used structure in commercial real estate waterfalls in the institutional world, and the one most frequently seen in crowdfunded syndications, is where the sponsor contributes 10 percent and investors contribute 90 percent, and all investors, including the sponsor, are treated equally (*pari passu*). As the sponsor is treated equally, there is a higher pref and a bigger split of profits to investors.

In this most common of structures, all parties receive profits until they reach an 8 percent preferred return. After the pref, the investors and the sponsor receive all their capital back. Finally, any realized profits are split 80/20—80 percent to investors and 20 percent to the sponsor (Figure 8.4). In this case, since the sponsor and investor are on equal footing, there is greater risk to the investor than in the prior two examples, but this is offset by a higher preferred return and a bigger split of profits.

Multi-tiered waterfall

To give you a better sense of the structure of a large institutional deal, also found in some crowdfunded syndications, let's examine a multi-tiered waterfall. We have the same profit split as before: 80/20. The sponsor co-invests 10 percent of the equity and the investors contribute 90 percent.

Distributions are as follows: All the investors, including the sponsor, receive distributions until they reach their pro rata share of the pref. Investors and sponsors then receive their capital back. Finally, the profit split is 80 percent to the investors and 20 percent to the sponsor.

After the initial 80/20 profit split and once investor returns hit a 15 percent IRR, the profit split switches to 60/40—60 percent to investors and 40 percent to the sponsor. If investor returns reach a 20 percent IRR, the split of all remaining profit goes to 50/50 (Figure 8.5). The idea behind this kind of structure is that the sponsor is heavily

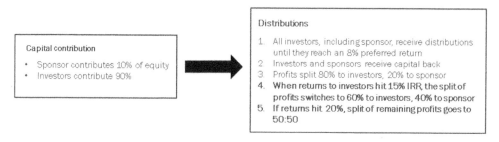

Figure 8.5 Multi-tiered waterfall structure

incentivized to outperform for the investors in terms of returns, because the better the deal performs, the more the sponsor makes.

Multilayered waterfall structure (a waterfall on top of a waterfall)

Now, let's examine a multilayered waterfall structure. For example, a private equity fund raises capital from pension funds (or endowments, or sovereign wealth funds, or whatever the source may be) and are paid a flat 2 percent fee for managing the money raised, plus they receive a promote. This may be a 2 percent assets under management (AUM) fee plus a 20 percent promote.[17]

In this kind of structure, the fund itself doesn't actually do the deals, it seeks sponsors to invest with and to manage. Its job is finding sponsors who have already found good deals and to then finance those sponsors' deals.

Such a fund might also give a promote type of waterfall structure to the sponsor. Here's how it might look.

1. The private equity fund raises capital from institutions, is paid a 2 percent fee for capital under management, and receives a 20 percent promote.
2. The fund then seeks sponsors to invest in, who in turn get a 20 percent promote incentive.

In this type of structure, there are now two tiers of promotes, two tiers of fees, and perhaps two tiers of preferred returns all baked into the same deal.

Waterfall insights

I have conducted extensive research and written extensively about waterfalls, including a detailed white paper that examines how their structures impact investor returns.[18] Here are some of the key findings of that work.

• The most common preferred return used in the industry is 8 percent—used by 40 percent of sponsors.
• The next most common is 10 percent, used by 30 percent of sponsors, and this is followed by 7 percent in 8 percent of projects. Next are 9 percent and 12 percent.

The remaining prefs range between 2 percent and 22 percent, with 0 percent rare but also seen.

- The most common hurdle used in multi-tiered waterfalls is the IRR (80–85 percent of all deals).
- The most common splits are either 90/10 or 80/20.

One of the most important findings is that waterfall calculations can be extremely complicated and are prone to error. They are traditionally calculated using spreadsheet software like Microsoft Excel. However, these have to be built manually, using complicated formulae. The more complicated the structure, the more likely a single error in the spreadsheet could throw off the calculations. The most complicated waterfalls involve multiple classes of investors, where each class has its own distinct set of waterfall terms. Class A, for example, might receive a 10 percent pref to a 90/10 profit split, but Class B might be getting an 8 percent pref to an 80/20 split. That's all well and good, but how do you actually calculate that? Add to it the possibility of multiple tiers and you end up with very complicated structures.

Another hazard for investors and sponsors alike is the use of the IRR to calculate waterfall distributions.[19] To calculate the IRR, a terminal value is needed; but during the life cycle of a project, there is no terminal value—so the only way to calculate an IRR is by estimating the terminal value. This begs the question, who gets to make that estimate and on what basis is it made?[20]

As an investor in a crowdfunded real estate syndication, make sure you are aware of the various promotes and related tiers that may exist within a deal. Make sure that whatever you're looking at is consistent with other deals you're seeing. Determine the going market rate for preferred returns and profit splits by comparing multiple similar opportunities. If you find something that seems off-kilter with all the other deals you see, and it is inconsistent with what you're generally seeing in the market, it may be a deal you want to pass on.

Most importantly, you always want to make sure that regardless of the promote structure, regardless of the preferred return structure, and regardless of the waterfall structure, your interests are aligned with the sponsor's—so that you get paid. In other words, be sure that you and the sponsor are ultimately both on the same page.

Financial key 7: Promote and fees

Referred to in other industries as a 'carried interest,' the promote is an incentive bonus paid to sponsors to compensate them for putting deals together and for executing on them. Specifically, it is an extra, disproportionate share of the upside from the transaction. It is part of what the sponsor is paid and is designed to motivate them to make money for everyone involved.

The promote

In a standard deal structure, the sponsor contributes some of their own money, the co-invest, as a portion of the equity to the deal. For example, the sponsor might contribute 10 percent and investors then contribute the rest of the equity. That portion of equity invested alongside investors is typically treated the same as the investors' contribution—that is, the sponsor takes on the roles of both developer and investor in the deal.

Once the deal is done, the bank first gets paid first. Then, a preferred return is paid in proportion to all investors, including the sponsor. Next, the invested capital is paid back to investors, and, finally, profits are split according to the agreed-upon promote; for example, the sponsor receives 20 percent, and the investors receive 80 percent. This is how the payout looks in order of priority: This 20 percent portion of the remaining money available for distribution (the profit) is the sponsor's promote. They receive this after the bank gets repaid and after the preferred return and invested capital have been paid back.

As stated above, the promote serves to incentivize the sponsor to do a good job for the other investors in the deal, and a properly structured promote contributes to aligning interests of the sponsor with their investors.

How sponsors earn their promote

The promote is a standard structure that's been used for decades in commercial real estate. But what exactly does a sponsor do to earn that promote?

Primarily, sponsors find deals that make the most financial sense. They operate in a hypercompetitive market to purchase properties, spending many hours weeding through hundreds of deals to choose the best opportunities and unlock their hidden value. Once located, the sponsor conducts thorough due diligence, and then they raise debt and equity to finance the deal. In some cases, they sign on the debt, which is a big responsibility—guaranteeing debt is not a trivial function.

After purchasing a property, the sponsor executes on their business plan, doing the actual construction or renovation and leasing up the property. When a professional sponsor has a high level of integrity, they have investor interests in mind at all times. This includes taking responsibility for maintaining compliance with all zoning regulations, investor regulations, accounting practices, securities laws, and other major concerns. Upon completion of a project, the sponsor sells the property for the highest amount and distributes the proceeds, including the profits.

In a well-structured deal, the sponsor does not make money until you, the investor, get your money back with interest (preferred returns), and they don't receive their agreed-upon promote until you get paid. This is what drives them and incentivizes them to do a good job for you.

Sponsor fees

However, sponsors can't do all this work solely based on the chance that they'll make money at the end of the day. That would be a misalignment of interests. Traditionally, sponsors also earn fees, and how those are structured can also show whether or not interests are aligned, as well as serving as a good indicator of the sponsor's integrity.

Sponsor fees are an incredibly important item to consider. Without a doubt, they are justified in a deal. You have to pay fees. In fact, you should want to pay fees. After all, as a passive investor you're paying somebody to do all the work that you don't have to do. However, you don't want to motivate the sponsor to only work for fees. Most particularly, you don't want the sponsor to make all the money they want to make *before* you make your money, which is something that can happen if fees are inflated unreasonably.

Promote versus fees

The main distinction between a sponsor's promote and their fees is as follows.

*   The promote is paid upon the deal being successful; it is only paid to the sponsor once the project has performed successfully.
*   Sponsor fees are paid during the course of managing the deal. These are paid before profits are calculated, and charged as a cost to the project.

Unlike the promote, fees are paid at different stages of the project, whether or not the project is proceeding as planned. These fees are often tied to specific tasks the sponsor must perform—so not tied to actual performance, but in compensation for services provided.

One thing to consider is whether the proposed deal fees are consistent with market norms. If they're not consistent, they can effectively eliminate the alignment of interests otherwise found with the promote and the structure of the waterfall. Yet their absence should not disincentivize that sponsor from doing the required work.

How fees are charged

Fees are charged either as a flat rate for a particular task that's involved, as a percentage of some other cost, or as a percentage of a line item in the project budget. In some cases, the fees charged to a project are those that a sponsor would otherwise pay to a third party.

For example, a sponsor might have an in-house architectural department, so the standard architect rates for doing a certain part of the work are a legitimate charge to the project. Since it is less expensive to have an in-house architect, it is not unreasonable that the sponsor can earn a profit by providing those services as long as they are charging standard market rates for those particular services.

However, while some fees are necessary to the success of a project, some are not. Architectural or engineering fees are necessary in any value add or ground up project. Regardless of whether the sponsor is supplying this line item or hiring a third-party firm to do the work, it's an inevitable and legitimate cost. Another category of fees, developer-specific fees, are more arbitrary.

For example, if a developer acts as the broker in buying and selling their properties, there's a legitimate case for charging a fee. This fee would need to be the same as a third-party broker might charge. Similarly with leasing fees, if the sponsor has an in-house broker, they can charge brokerage fees for leasing activities. An exception to this is if the sponsor charges an acquisition fee on top of paying an outside broker. You may see a situation where brokers are used on both buyer and seller sides in a transaction, and the developer is charging an acquisition fee on top of that. There may be a disposition fee also built into the operating agreement for the project, that awards a developer a fee on top of a fee to a third-party broker.

It is in these kinds of discretionary fees that you can find excesses. Additional fees on top of regular brokerage fees that a sponsor might charge a project are often casually dismissed as being 'just to keep the lights on.' That may be so, but I've seen 1–2 percent fees charged on $30 million projects, and that speaks to some expensive electricity bills.

Think of a scenario where a project fails to make a profit and the developer has to sell in distress, losing you part or all of your equity investment. You probably would not want to see the developer taking a percentage-based fee from the sale of the property, benefiting them while further exacerbating your losses. As with all fees, most on their own are justifiable, but in aggregate—meaning the sponsor is layering too many fees on to a project—they can be egregious and something you want to be on the lookout for. You don't want a sponsor to be motivated by fees, because that sets the stage for a conflict of interest between you and them.

The next sections include some more examples of what to keep an eye out for and show the distinctions between market rate and developer-based fees.

Market rate fees

As I mentioned, when the sponsor has an in-house architect, the market-rate fee for architectural drawings is legitimate. The same goes for general construction activities. The sponsor can charge these things to the project at the same rates as a general contractor would. The sponsor may also charge fees for asset or property management. If the sponsor manages the project after its completion, they could hire a third party to manage the property—collecting rents, dealing with tenants, and doing other management tasks. However, if they have an in-house asset manager, they can charge market-rate asset management fees to the project.

In sum, the market-rate fees that a developer charges a project by performing a critical function in-house are fees that must be paid by the project to somebody, be it a third party company or the developer. Depending on the asset type and investment strategy, market-rate fees include things such as:

- brokerage, buying and selling
- leasing fees
- architectural drawings
- general construction
- property management

Developer-specific fees

In addition to any market-rate fees that third parties might charge, there are developer-specific fees that you need to consider, like the acquisition fee on top of a brokerage fee and disposition fees upon the sale of the property. A sponsor may charge a property management fee, development fee, or construction management fee, all on top of what the project is already paying to a third party at market rates.

There could also be an AUM (assets under management) fee to manage your money. This typically ranges between 1 and 2 percent. For example, if there's $100 million in equity raised, they may just charge a flat 1 percent—$1 million a year—to manage that money. There may be finance fees for arranging the project's debt, and a sponsor may charge a fee to the project for guaranteeing the debt.

These are all common developer-specific fees that you might see in a deal. Since they sit over and above what is absolutely required for a project, these are key fees to keep an eye on:

- acquisition fees
- disposition fees
- property management fees
- development fees

- construction management fees
- AUM fees
- finance fees
- guarantee fees

Acquisition and disposition fees

Acquisition and disposition fees are developer-specific fees that a project might pay to a developer. These fees are above what is paid to an outside broker upon the acquisition and sale of a property; that is, in addition to the market-rate brokerage fees, they may also charge this acquisition and disposition fee. Typically, if a sponsor charges an acquisition/disposition fee, it can be as high as 2 percent 'in and out,' which means 2 percent for the acquisition and 2 percent upon sale of the property.

Rather than just 'keeping the lights on,' some sponsors may charge the acquisition/disposition fee to the project to defray what are called 'dead deal' costs. For a sponsor, these are costs associated with searching for, underwriting, and rejecting multiple deals before actually finding a winner. The acquisition/disposition fee, though an additional cost, can work in the investors' favor because it can keep the sponsor from jumping at the first deal they find, just to get into the business of developing.

An effective sponsor must have the patience to sort through and screen out all the dead or bad deals. In the end, they only bring the best deals to you, the investor. Therefore, this acquisition/disposition fee compensates the sponsor for the out-of-pocket expenses associated with separating the wheat from the chaff to bring you just the winners.

Property management fees

Another developer-specific fee is the property management fee. Often, the sponsor hires a third-party property management company to manage a rental property. Their services include day-to-day operations, leasing, maintenance, accounting, etc. As with the brokerage function, the developer may have their own property management team or may work with a subsidiary that charges market rates.

This type of fee is typically 3 to 5 percent of gross revenues, usually paid monthly, plus the cost of any out-of-pocket expenses. If the sponsor manages a building and they need to fix the roof, that's an out-of-pocket expense they charge to the project. In addition, they might charge a management fee on top of that, charged at the same rate (3 to 5 percent of the cost) for handling that specific problem.

Development fee

The sponsor might also charge a development fee to cover broader development functions, such as the early due diligence stage and all the processing of entitlements for a project. These are time-consuming processes for which the sponsor may seek compensation.

The development fee also covers pre-construction due diligence items, such as environmental testing, traffic reports, compliance with local regulations, planning petitions, and other similar costs.

Development fees also include such line items as general entitlement and permitting work, including managing all third-party professionals hired for the projects, overseeing architects, engineers, contractors, soil testing, entitlement expediters, etc. Overall, a sponsor may charge a general development fee for a number of functions related to developing a property. Typically, sponsors charge up to 5 percent of the total cost of the project to compensate them for this work. Alternatively, they might allocate this fee as a percentage of either soft or hard costs only. On rare occasions, it is a flat fee.

Construction management fee

If the deal has a large construction component, such as a ground up project might have, or is a very heavy adaptive reuse or redevelopment project, the sponsor may charge construction management fees to cover their costs of managing the physical construction process. As discussed earlier, they may charge market-rate fees for certain in-house construction-related functions. This fee can cover architectural and other professional services, the management of general contractors or subcontractors, or project management. It also covers bidding costs, budget control, and the sponsor's overhead for payment approval and processing.

The construction management fee can also cover the cost of design and change orders. Nearly all projects have a fairly constant flow of design changes, or other changes that arise when unexpected site conditions are discovered. All of this requires considerable time and expertise to manage. This fee compensates the sponsor for the expertise, resources, costs, and time spent dealing with these issues.

In addition, this fee also covers ongoing decision-making activities, such as choosing interior finishes for a project or handling inconsistencies between construction documents and actual development site conditions. Throughout my career, I've never seen a set of architectural plans that didn't contain some inconsistency somewhere. For example, during construction, it might be discovered that a support feature juts out into a hallway. The developer must determine what changes need to be made to correct the issue, and they will need to do it within code and on the fly to minimize cost overruns.

The construction management fee can also cover site inspections, cleanup, and the completion process paperwork, including whatever documentation is involved in managing the construction component of a deal.

Construction management fees can be charged up to 5 percent of total costs, depending on the extent of services the sponsor provides. The fee can also be charged as a percentage of only the soft or hard costs or as a percentage of both, or the sponsor may charge a flat fee, similar to the development fee.

Assets under management fees

AUM fees are more commonly associated with private equity funds that manage large sums of institutional capital. In those instances, there is a dedicated team of professionals who find sponsors in whom to invest. In the past, this type of fee was charged at 2 percent of the total AUM. However, during the recession of 2007–2008, the rate decreased. Today, it is typically 1 to 2 percent of the total AUM, depending on the size of the fund.

For example, let's say a fund has $100 million of AUM or invested capital under management. In this case, the management team will earn $1 million to $2 million a year. This AUM fee covers all the overhead incurred on behalf of investors to manage the process.

The AUM fee covers general overhead that may include strategic decision-making, including allocating capital expenditures or deciding upon which deals to approve. Private equity funds might have investment committees whose expenses are covered by this fee.

Furthermore, the AUM fee covers the cost of raising funds in the first place. If the fund has a management team working to raise $100 million, the fund charges an AUM fee to cover the costs associated with these activities. The fee can also cover dead deal costs. Keep in mind that if you see an AUM fee, you should ensure there's no double-dipping for other line item costs or fees in the deal.

Financing fees

Sometimes, a developer charges a fee for engaging in the process of putting together debt financing, which includes working with numerous banks, negotiating the best rates, and securing the financing. Although banks also charge fees for putting loans together, it is not uncommon or unreasonable for the sponsor to also charge a fee as compensation for arranging the financing. Typically, debt financing fees are around 1 percent. For example, the fee for a $10 million loan would be $100,000.

Guarantee fees

For a guarantee, the fee is usually around 2 percent, which is calculated on the amount of the guarantee. Sometimes, a loan may require a personal guarantee on the entire amount, or it may only require a personal guarantee for the top 10 or 20 percent of the total loan amount. The sponsor's fee then would be a percentage of what they guarantee.

How fees can impact investor returns

Developers fees and the influence they can have on a project, particularly in aligning interests, is important, so to help you evaluate deals you come across, let's walk through a fictitious deal step by step.

In our fictitious deal, the total cost of the project, including all hard and soft costs, is $20 million. Though high, for the sake of this example, let's assume a bank will provide 85 percent of the total cost, leaving $3 million of equity required. The sponsor contributes 10 percent of that, so $300,000, and investors bring in the remaining $2.7 million.

The deal timeline works like this. Using their own and investor equity, the sponsor buys the development land for $3 million at the outset and builds the following year—call it Year 1. From Year 2 the property generates operating profit—NOI—and in Year 5 the sponsor sells the project at a 6 cap for almost $21.7 million (Table 8.1).

Table 8.1 Impact of fees on investor returns

	Year 1	Year 2	Year 3	Year 4	Year 5
Buy land	$ (3,000,000)	$ -	$ -	$ -	$ -	$ -
Build	$ -	$(17,000,000)	$ -	$ -	$ -	$ -
NOI	$ -	$ -	$ 600,000	$ 1,200,000	$ 1,250,000	$ 1,300,000
Sell Building	$ -	$ -	$ -	$ -	$ -	$ 21,666,667
TOTAL	$ (3,000,000)	$(17,000,000)	$ 600,000	$ 1,200,000	$ 1,250,000	$ 22,966,667

Table 8.2 Cash flow

| | No fees | | |
	Total profit	IRR	EMx
Investor	$ 4,548,000	22%	2.68
Sponsor	$ 1,768,667	43%	5.90

Table 8.3 Deal with no fees

Acquisition	2%	$ 60,000
Disposition	2%	$ 433,333
Property management	5%	$ 217,500
Development fee	5%	$ 850,000
Construction management	5%	$ 850,000
AUM	2%	$ 54,000
Finance fee	1%	$ 170,000
Guaranty fee	2%	$ 340,000
TOTAL FEES		$ 2,974,833

For the purposes of this example, the investors are paid an 8 percent preferred return and the sponsor receives a 20 percent promote. And as we are looking at the impact of fees on a project's distributions, let's also assume that the sponsor charges the project no fees in this scenario.

Table 8.2 shows how the cash flows appear at the project level in this scenario.

After the preferred return has been distributed, and the sponsor has received his share of the profits through his promote, the investors will have received back a total of $7,248,000, including the return of their original capital, which equates to $4,548,000 of profit. This represents a 2.68× equity multiple on their original investment and a 22 percent IRR.

The sponsor, who only puts in a co-invest, has a higher IRR, driven largely by the promote—hence this idea of a 'disproportional' share of the profits, or what incentivizes sponsors to outperform. In terms of total profit, the sponsor receives $1.768 million, which, compared to the equity investment, has a higher IRR and a higher equity multiple relative to their 10 percent co-invest, but is a significantly lower proportional share of the profits—as you would expect.

The example in Table 8.2 shows how the deal plays out without fees charged to the project. However, should the sponsor decide to charge every single one of the different line item fees—a 2 percent acquisition fee, a 2 percent disposition fee, a 5 percent property management fee, a 5 percent development fee, a 5 percent construction management fee, a 2 percent AUM fee, a 1 percent finance fee, and a 2 percent guarantee fee—the deal will look very different (see Table 8.3). The sponsor would be paid nearly $3 million in total fees before anything is distributed to the investors.

Table 8.4 Comparison of all-fees and no-fees scenarios

| | No fees | | | | All fees | | |
	Total profit	IRR	EM×		Total profit	IRR	EM×
Investor	$ 4,548,000	22%	2.68	$ 2,406,120		14%	1.89
Sponsor	$ 1,768,667	43%	5.90	$ 3,610,547		67%	13.04

Remember, before investors are paid a penny, the sponsor takes all of these fees. So let's look at how fees affect the difference. Table 8.4 compares the no-fee scenario with the all-fees scenario.

In the no-fee scenario, the total profit is the number to focus on. In this case, the investors get a total profit of $4,548,000. On the other hand, in the all-fees scenario, the total profit collapses to $2,406,120 after taking out fees. As you can see, the sponsor's return goes from $1,768,667 to $3,610,547—around $2 million in fees goes from the investor to the sponsor.

In any deal, you're looking for balance. From these examples, you can see that sponsor fees can be detrimental to alignment of interests, especially since they're being taken out before investors even get their money back—forget earning a profit. In other words, it is investors' capital that pays the sponsor whether a project is successful or not.

Evaluating if a sponsor's fee structure is reasonable depends on how the sponsor company is structured. A strong, vertically integrated sponsor, for example, may charge more fees and a smaller share of the profits on the back end. If the sponsor has a brokerage function, an architect in-house, a management function, or whatever else, he might be more inclined to charge fees to the project to cover his overhead. Yet on the back end, he may offer you, the investor, more in terms of profit split.

On the other hand, a smaller, newer developer may want to demonstrate his confidence in the deal by taking lower fees but asking for more on the promote. This sponsor might think that while they are performing all these functions, they won't charge fees for them. Instead, they'll show investors that they are confident in the deal by stacking their returns more heavily on the back end performance of the deal.

Identifying what fees are charged

Some sponsors display fees prominently in their offering documents or even in their pitch decks. However, more often than not, they're buried somewhere in a 200-page set of offering documents. If the sponsor fees are not summarized front and center, be sure to review the documentation, create your own summary, and compare the fees for this particular deal with fees in other deals you're considering.

Don't be afraid to ask the sponsor what certain fees are for and exactly how they're calculated. In fact, don't be afraid to ask a sponsor to explicitly name all the fees they charge and what those fees go towards. Never be afraid to ask the fee question; it's very important.

Below is a list of fees and reimbursements that were buried deep inside the contract of a deal I reviewed recently.

Fees and Reimbursements

Section 6.1—Acquisition Fees
The Company shall pay the Investment Manager an Acquisition Fee equal to 3 percent of the total projects costs for each Property acquired.

Section 6.2—Disposition Fee
The Company shall pay the Investment Manager a Disposition Fee equal to 3 percent of the total projects costs for each Property sold.

Section 6.3—Property Management Fee
The Investment Manager shall be paid a monthly management fee equal to five percent (5 percent) of the gross monthly revenue of the Properties directly from each Property account.

Section 6.4—Reimbursable Expenses
The Investment Manager's reimbursable expenses shall be paid directly from each Property account.

Section 6.5—Leasing Commissions
The Investment Manager shall be paid market rate leasing commissions in connection with each new lease procured by the Investment Manager (or renewal of existing leases) during the term of this Agreement.

Section 6.6—Development Fee
The Investment Manager shall be paid a development fee equal to fifteen percent (15 percent) of all hard and soft expenditures in connection with any development, redevelopment, refurbishment, renovation or improvements made to the Properties.

Section 6.7—Finance Fee
The Investment Manager shall be paid a fee equal to 1 percent of the loan amount for any loan obtained for the Properties.

Section 6.8—Guarantee Fee
The Investment Manager shall be paid a fee equal to 1 percent of the loan amount for any loan obtained for the Properties that requires a personal guarantee of repayment and/or if the lender has recourse against the Investment Manager or the principals of the Investment Manager.

As you can see, the list is quite long, and not only are certain line items egregious, but in total they are excessive too.

First of all, the Company shall pay the investment manager an acquisition fee of 3 percent. That's 3 percent of the total project cost for each property acquired. Investors then pay a disposition fee equal to 3 percent of the gross sale price of the property. That's a total of 6 percent off the top whether the deal performs well or not—and in this case the sponsor get the fees even if investors lose money.

The investment manager in this deal is paid a monthly management fee equal to 5 percent of gross monthly revenues. On top of that, their reimbursable expenses are also paid.

Regarding development fees, the investment manager is paid a development fee equal to 15 percent of all hard and soft expenditures in connection with any development. For the finance fee, the sponsor gets 1 percent of the loan amount for any loan obtained for the property, as well as a guarantee fee (also 1 percent) for any loan amount that requires a personal guarantee of repayment.

Keep in mind, this is a real-life example of sponsor fees, so it hits home the point that you must look for the sponsor fees and decide whether they're reasonable compared to other deals you're considering. Do they seem fair? Are you comfortable with the fees?

Incidentally, in this same contract, there was an additional clause saying that "[a]n appropriately allocable portion of all salary, payroll taxes, training expenses, worker's comp insurance and other employment costs incurred" by the sponsor in connection with the properties shall be reimbursed to the sponsor by the investors, or by the project. What is an "appropriately allocable" amount? When combined with all the fees on this deal, it's clear the sponsor is motivated by fees and there is a misalignment of interest. You should walk away unless you can renegotiate the terms.

As an investor, these are things you need to look for. If you find them, ask the sponsor for clarification. If you're not satisfied with the response, move on to another deal. If you are satisfied, then go ahead and finish up the remainder of your due diligence.

As a small individual investor in a large deal, you carry little or no negotiating power. An institutional investor, who can write the kind of checks that really make the difference on whether a sponsor moves forward on projects, has a lot more power. These big investors refuse to let a sponsor receive a single penny until all the investment capital has been returned with, at minimum, the preferred return. These investors want to see the sponsor motivated to outperform by ensuring they are not motivated by fees up front.

While you don't have the negotiating leverage to influence a deal this way, you need to be sure there is a continued alignment of interests at all times. Ultimately, you must rely on your judgment and sense of what is fair in any particular deal. Look at dozens of deals and review their fee structures. You'll start to get a sense of what is reasonable and what is egregious.

Financial key 8: Leverage

Let's move on to one of my favorite subjects: leverage. Leverage is the use of borrowed funds, usually from a bank, but sometimes from private lenders. Some common terms for leverage include 'debt,' 'mortgage,' or 'loan.' The subject of debt could be the single most important topic for you to understand, because, like most things in life, too much of a good thing (which debt can be) can be very bad for you.

When extending credit to a borrower, lenders typically secure their loan by holding the subject property as collateral for the loan. In the event that the borrower fails to pay or to make timely payments to the lender, *the lender reserves the right to take ownership of the property away from the borrower* in order to recover the outstanding loan by selling the property on the open market. Foreclosure, or the lender's right to take ownership of a property if a borrower fails to make payments, is key to how lenders secure loans and how they ultimately wield the greatest power in any deal.

Why take the risk?

Of course, this begs the question: If a lender can take ownership away from a borrower, why would you use leverage? If a sponsor runs the risk of losing the property to a lender, why take that risk? Well, first of all, financing a deal with all cash is usually impractical; if you want to buy a $10 million deal, it's unlikely that you or the sponsor has $10 million available in cash. Using only cash depletes capital resources quickly, which means a sponsor can do fewer deals, leading, in turn, to elevated risk concentration due to lack of diversification.

Borrowing can help make an investment more profitable, as you'll see, because borrowing helps amplify investor returns (hence the term 'leverage') while reducing risk (to a point) by lowering financial exposure. In other words, if you put less of your own money into a deal, then you obviously have less money at risk. However, because borrowing can also increase risk by making a deal vulnerable to financial failure and foreclosure, there's a balance between how much capital you want to risk and how much you want to increase the risk of losing your own capital by taking on too much debt.

Layers of debt

Leverage can include multiple layers of debt in what becomes a cascading order of priority, as we've discussed in the chapter on waterfalls but that I'll add to now to demonstrate the power lenders wield over a project. Similar to the example we described in the waterfall section, when you buy a house, you have to put some money down. When you make a down payment, the bank then gives you a loan and might also provide you with some kind of home equity line of credit (HELOC).

If you sell the house, there is a certain order in which everyone is paid back—similar to the waterfall—with the bank mortgage being paid back first. After the mortgage, the bank is paid back whatever remains owed on the HELOC. Finally, you are paid back. So, the cascading order of priority for the debt is:

1. first to the bank with the mortgage
2. second to the bank with the HELOC
3. finally to you

Similarly, when you invest in commercial real estate, multiple layers of finance might be placed on a deal. These layers include a first-position loan, which will be paid off first. A second-position loan may exist, and this is paid off second. Then, a third-, and fourth-position loan, etc. Then, a layer of mezzanine debt—or more commonly these days, preferred equity[21]—may exist, which is typically the last layer of finance before you come in with your equity, or your investment.

This structure is called the 'capital stack,' where multiple layers of money are layered in a specific order of priority. These layers resemble a stack of money, hence the term 'capital stack.' Figure 8.6 illustrates the capital stack.

First-position loan

Starting at the very bottom of this money stack is bank debt—the first-position mortgage. This is usually the largest part of the capital stack. It's typically provided by a government-regulated bank. At the same time, it's also the lowest-risk part and has the lowest interest

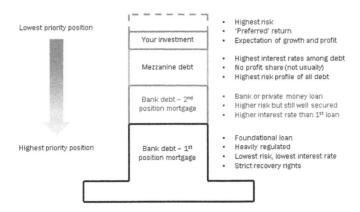

Figure 8.6 The capital stack

rate. The bank debt includes very strict recovery rights. If a borrower fails to pay, the bank has clear regulatory guidance and requirements on precisely what they can do, and when, to recover their capital and is, as a consequence, the least flexible component of the capital stack.

The bank debt sits at the bottom of the capital stack. In this way, it is foundational, like the concrete foundation that forms the strongest part of a building.

Second-position loan

As you climb up from this foundation, there may be another layer of bank debt. In theory there could be any number of layers in a capital stack, but for this example, we have a second layer of bank debt—the second-position mortgage. This might be a loan from a bank or a private lender. It carries higher risk than the first-position loan at the foundation but is still well secured. And it commands a higher interest rate because of the slightly higher level of risk than the first-position loan.

Mezzanine debt

Continuing upward, mezzanine debt (sometimes called 'mezz') or preferred equity sits above the second-position loan. Mezzanine debt commands the highest interest rate of all debt in the capital stack, because it carries the highest risk.

There's no profit share for anything that comes into a deal in the form of debt including mezz debt or preferred equity, just an interest payment and the use of the real estate or operating entity respectively, as collateral.

Equity (your investment)

Above the various debt components, at the top of the capital stack sits your investment. You, the investor, take on the highest risk. In exchange for taking on that that risk, you get a preferred return and a share of the profits. Looking at Figure 8.6, the top position—your equity investment—represents the lowest priority, while the lowest position of the capital stack—the first-position bank debt—signifies the highest priority.

Priority position

What exactly does priority position mean? If a borrower fails to satisfy the terms of a mortgage and defaults on the loan by failing to make a payment, then the lender can foreclose. In foreclosure, the first-position lender can take ownership away from the borrower. Indeed, if the borrower fails to pay any of the debt positions in the capital stack, whoever they fail to pay will have the right to take the ownership away from the borrower.

This is how that works: If a borrower stops paying on the first-position bank debt, the first-position lender can foreclose on the property, wiping out all other positions in the capital stack. This 'wipe out' means these other parties can lose everything with no claim to recoup their losses, as illustrated in Figure 8.7.

Lenders don't want to foreclose; they don't want to take possession of real estate, because they are not in the real estate business—they're in the banking business. However, they have a fiduciary responsibility to their shareholders, and as federally

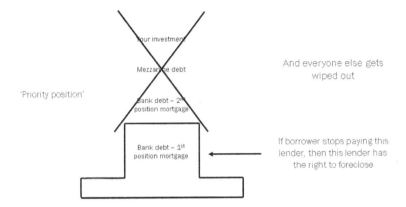

Figure 8.7 Failure to pay first-position debt

regulated entities, they are bound by regulations that mandate, under specifically defined conditions, that they must foreclose or take other steps to recover everything they are owed. However, when they do foreclose, they only need to sell the property for the amount that they loaned in order to recover everything. They have no obligation to other parties in the capital stack. If the borrower stops paying the first-position mortgage, the lender forecloses, takes back the ownership of the property, and sells it on the open market.

Essentially this means that you, the investor, and everybody else in the capital stack are toast. The lender's foreclosure will wipe you out. This reflects what it means to be in a priority position: The first-position bank lender sits in the highest position of priority for repayment and holds the power to wipe out all other parties in the capital stack, should the borrower default on the first-position loan.

Failing to pay second-position debt

In some circumstances, a borrower may stop paying a second-position lender before they stop paying the first. In such a situation, the second-position lender now has the right to foreclose on the property and take ownership away from the borrower. However, in order to continue to own the property, the second-position lender must now take on all the responsibilities the borrower originally held. This means they must continue to pay the first-position bank debt. In order to maintain ownership of the property, they have to keep that, and any other loans below them in the capital stack, current.

As shown in Figure 8.8, in this kind of situation, anyone above the second-position lender on the capital stack gets wiped out, which includes the investors. As long as the second-position lender keeps the first-position loan current, they can take ownership of the building away from the borrower, and you, the investor, lose everything.

The second-position lender is now in the position to sell the building if they want or to execute on the original borrower's business plan. They are now the new owner so can do whatever they want to do.[22]

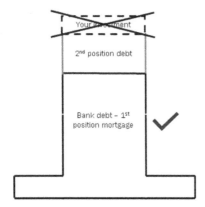

Figure 8.8 Failure to pay second-position debt

Understanding positions in the capital stack

It's essential that you know who is ahead of you in the capital stack, who holds higher priority, and what rights they have, under what conditions, to take the property back from the borrower and wipe you out. Furthermore, if the developer fails to perform as promised, you also need to understand your own rights. What rights do you have in those circumstances, and what are your options?

In some circumstances, you may have some foreclosure rights, or you may have the right to eject the developer from the project. In any event, you need to consult the offering documents to identify what rights you do have in the event the borrower or sponsor fails to perform.

In addition, different investor classes often hold different positions of priority over each other. These are not so common in crowdfunded syndications, but when they do occur, they make for among the most complicated of structures and the hardest for sponsors to manage accurately (it is notoriously difficult to calculate waterfall splits in structures with multiple investor classes). For example, the investor level may be split between a Class A and a Class B structure, where each has a different preferred return, different hurdles, and different promotes (see Figure 8.9). In such a structure, one class of investors may also have priority over another class. You may even find that one class of investor could have the right to wipe you out, or vice versa—though these rights don't generally emerge until the middle of a deal's life cycle if things have gone wrong and the sponsor has had to take in new investors to make up for capital shortfalls.[23]

It is imperative that you pay attention to these details very closely before making an investment. That's not to say that you shouldn't accept being in the lowest priority position—it is, after all, inherent in being an investor in a real estate deal. But when entering into a deal, you should at least make certain you know your position in the capital stack and what rights you have. Should things go south, if the borrower, or sponsor, stops performing as promised, know in advance what rights and remedies you have.

Once you've considered your rights, relative to the other parties in the capital stack, check to see how that compares against other deals you're considering. You should compare whether a particular deal is consistent with standard market terms. If the deal seems

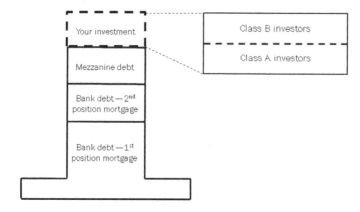

Figure 8.9 Your investment: Class A and B investors

to be in alignment with everything you're seeing, that's a good sign. If it's completely out of whack, it may warrant further investigation on your part to find out just why it has certain terms or conditions.

How leverage works

There is, of course, a flip side to ceding power to a lender in a deal: having debt and leveraging a deal can help investors make a lot of money. Leverage is typically used to increase yield to the investor. To be accretive to a deal, the interest rate on the leverage must be lower than the unlevered yield on the investment. We'll take a look at this in a moment.

The key to understanding the benefits of leverage is that while it can increase returns, if it is overused, it can also expose a deal to excess risk; while leverage can make good things go better it can also make bad things worse by exposing the project to undue risk of foreclosure.

Table 8.5 provides an example of the impact of leverage on overall profit. Three scenarios are shown.

Scenario 1. A deal with no leverage

In the zero-leverage situation, you buy a building for $1 million, pay all cash, and have a value increase of 10 percent. In this scenario, the total profit is $100,000, a 10 percent return on that $1 million purchase.

Scenario 2. A deal with 50 percent leverage

In the second scenario, you leverage (or borrow) 50 percent. This means that with your $1 million, you can buy one $2 million building or two $1 million buildings. In other words, you borrow 50 percent—you borrow $1 million and add your $1 million.

In this scenario, the market value increases by 10 percent. So the new building (or buildings) is now worth $2.2 million. This gives you a total profit of $200,000, a 20 percent return on your $1 million investment with 50 percent leverage.

Table 8.5 Impact of leverage

Leverage	Invest	Value increases 10%	Total profit	Total return on $1M
0% leverage	Buy one building for $1M	Building now worth $1.1M	$100,000	10%
50% leverage Borrow $1M	Buy a $2M building or 2 $1M buildings	Building now worth $2.2M	$200,000	20%
75% leverage Borrow $3M	Buy a $4M building or 4 $1M buildings	Building now worth $4.4M	$400,000	40%

Scenario 3. A deal with 75 percent leverage

In a 75 percent leverage scenario, you use your $1 million to borrow $3 million. With this 75 percent leverage, you can either buy a $4 million building or four $1 million buildings.

Again, there's a 10 percent increase in value, making your total portfolio (one or multiple buildings) worth $4.4 million. This gives you a $400,000 profit, or a 40 percent return on the $1 million investment. In this scenario, the effect of leverage is to quadruple the returns you could make on that same initial $1 million investment.

While this is an example of how leverage impacts overall profit, taking on debt is not always beneficial. Taking a high-leverage, short-term loan at a low interest rate at the peak of the market, for example, can really cause problems. It's highly seductive to borrow 90 percent for 12 months or 24 months at a low interest rate. You might believe that's a good idea because you'll get a great return on the equity you put in. And because it's high leverage, you don't need to put as much capital into the investment that you might otherwise need to.

In the run-up to the 2007–2008 market collapse, this scenario occurred frequently. Here's what happened: Borrowers took on short-term loans at low interest rates with little invested money. When property values dropped significantly (below bank lending thresholds), borrowers seeking replacement loans to cover the earlier loans had to come up with higher down payments. However, due to the recession, vacancies increased, leaving borrowers without adequate cash.

But then the loans came due. Even borrowers who did have enough cash found themselves in a position where all their equity had been wiped out, leading them to wonder what the point would be of putting more money into such a deal; if a property is worth less than the debt owed, there is little point in investing more in it. Since they held secondary positions behind other equity, borrowers had zero incentive to add more money into the deal, so they defaulted. This set off the cascade of foreclosures. The vast number of foreclosures that occurred were in large part due to the excessive leverage borrowers had placed on commercial properties, because those properties had dropped so far in value they were no longer worth even the debt on them.

Yet high leverage isn't always bad. Let's say you've got an office building that is fully occupied by Google. You could take high leverage on that building, because in order for your tenant to stop paying you, they would first have to go bankrupt. The likelihood of Google going bankrupt any time in the foreseeable future is very low.

That said, even with high-credit tenants, there are risks. Think about companies like Sears, or Circuit City, or Blockbuster Video. Each of those was a high-credit tenant in their heyday. You could have put them into buildings, highly leveraged, and then suddenly found that they went bankrupt. Then you'd be stuck not only with an empty building, but also a stack of bank debt.

In general, you want to see lower leverage for lower-credit tenants, mainly because there is a higher risk of this kind of tenant not paying, going bankrupt, or moving on. Or you could have multi-tenant buildings with small tenants to spread the risk. If the economy goes down, while some may stop paying, go bankrupt, or move out, they represent a lower proportion of the total tenancy so have a lower impact on vacancy rates overall, and they should be easier spaces to fill during a downturn.

In this case of lower-credit tenants with smaller square footage leases, the banks will help sponsors maintain lower leverage—simply because they won't offer higher leverage with this kind of tenant mix. Banks want to see high-credit tenants with long leases, and when they do, they reward sponsors with higher leverage at lower rates. Paying attention to the terms a bank is lending on, relative to other buildings, will provide a good indication for you as to how they view the risk of an investment you are considering.

Appetite for risk

The takeaway here is that leverage is an important consideration in evaluating the risk in any investment. From a risk-return perspective, a 10 percent return on a low-leverage deal might be better than a 15 percent return on a highly levered deal. It really depends on your risk-return profile, or your appetite for risk.

You may prefer to take a lower return and only invest in deals with 50 percent leverage. If you're only putting 50 percent leverage on a deal, you reduce the potential yield and equity multiple. Even though you get a lower overall return, you also have less risk.

Since the 2007–2008 downturn, the change in the regulatory environment for lending actually presents an opportunity for you as an equity investor. Since banks are more conservative in their lending practices than they were pre recession, equity must pick up the slack. Sponsors need money. They need the equity. And since they can't access as much debt as they once could, there's a bigger market for equity out there than there was before the downturn. Before and during and recession of 2007–2008, crowdfunding commercial real estate syndications was prohibited. Now you have the opportunity to invest in deals that were previously off limits.

Different types of lender

Having a lender on a deal is a good sign that the deal has already been underwritten and approved by someone else before you even look at it. The lender will have already reviewed some key aspects of the deal. This doesn't mean the lender is infallible (they're not). Lenders do make mistakes. Presumably, they've covered at least some of the basics like ownership, questions of title, compliance and the like. They've looked at surveys, addressed environmental concerns, have conducted physical inspections, and have underwritten the sponsor's assumptions. In general, having a lender on a deal assures you that a credible third party has already critically examined the project and underwritten the numbers.

This does not abdicate you from the responsibility of double-checking all the work, or at least taking a close look. To paraphrase Mitt Romney's (in)famous words, banks are people too.[24] And banks make mistakes. Don't assume that because a bank has under-written a deal, everything's already been checked.

Specifically, the caliber of a lender can be an indicator of the credibility of the under-writing. A prominent Main Street bank probably has higher standards than a local bank. This could be an unfair statement, but it's a realistic indicator of the credibility of the deal.

However, the leverage a bank gives to a borrower depends on several factors that are not necessarily aligned with investor interests. For example, *ceteris paribus*, a lender might offer a sponsor a higher loan amount in return for a fee, thus lending more than other banks. This can distort the impression you get of the leverage and the bank's confidence in the deal.

In certain instances, the lender's philosophy may differ from your own. The bank might be particularly inclined to a certain type of real estate lending because some-body on the board is oriented to that type of property. The bottom line is that you don't know if the bank's lending practices are consistent with your own risk–return preferences. Nevertheless, the lender on a deal is a useful indicator for you when deciding whether or not sufficient groundwork has been laid for the due diligence you're going to be checking.

Relevant terms

Let's look at a couple of key terms that are relevant to leverage: loan-to-value (LTV) ratio and debt service coverage ratio (DSCR).

Loan-to-value ratio

LTV ratio is a calculation used by banks to measure their exposure to financial risk. It is the ratio of the amount of debt relative to the actual value of the asset in percentage terms. Default risk increases as the loan-to-value ratio increases. If the property's value is $1 million and the loan is $1 million, that is a LTV ratio of 100 percent, which represents the highest default risk that a bank could have on a loan—banks share in none of the upside and take on all of the downside, while the borrower has nothing to lose.

The LTV a lender offers depends on their appetite for risk as well as on governing regulations. A conservative LTV ratio might be 50 percent or less. An aggressive LTV ratio might be anything over 75 percent. If a sponsor borrows 75, 80, 85 percent or higher, there's a very aggressive lender behind that debt and a less conservative sponsor who is borrowing a lot to do the deal.

Of course, there are variations on these rules—specifically related to tenant credit, weighted average lease term, and distribution of tenant square footage relative to the overall size of the building—that banks consider. But the principal remains the same: The higher the lender perceives the risk, the lower the amount of principle they will be willing to lend and the higher the interest rate they will apply.

The calculation for LTV is the total mortgage loan balance—the total debt—divided by the purchase price of the property, or the value of the property.

$$Loan\text{-}to\text{-}value\ ratio = \frac{Mortgage\ loan\ balance\left(debt\right)}{Purchase\ price\left(or\ asset\ value\right)}$$

Interestingly, this ratio can change over time. Let's say a sponsor buys an office building for $10 million and borrows $7 million from the bank. That's a 70 percent LTV. If there's a market correction and rents come down or tenants move out, the property's value may decrease. In this example, let's say the property value dropped to $9 million. The bank debt is still $7 million, but the LTV ratio has increased from 70 percent to 77.8 percent.

In a scenario like this, there can be significant ramifications. In theory, as long as debt service is maintained (i.e. as long as the borrower repays the bank), the bank doesn't bother the sponsor, even if the LTV ratio increases. However, in times of deep recession, banks can actually call in a loan if they feel there's been a material adverse change of circumstances. During a recession, a lender may ask to assess the actual value of all the properties they've lent against and recalibrate their LTV balances. That's when they may discover that some loans may be out of sync with regulations.

A material adverse change most often applies to construction loans. In those loans, in order to complete construction on the building, borrowers during recessions may want to continue to draw down on a loan to which both they and the bank had committed. Though considered a controversial application of this concept, during the 2007–2008 recession banks in some cases refused to extend credit. They argued that as there had been a material adverse change and the value of the asset was lower than they had originally expected or predicted, they were no longer bound to continue to finance construction draws. Banks either terminated the loans or demanded that the borrowers bring in more equity to the deal until the LTV was back in compliance with the original loan terms.

Debt service coverage ratio

Another important banking term used in relation to bank debt is the debt service coverage ratio (DSCR). It is sometimes called 'debt coverage ratio' (DCR). DSCR is another ratio lenders use to assess whether or not they will make a loan, and if so, for what amount. For lenders, DSCR is also a measure of risk. It measures how much NOI a building generates relative to the cost of paying the mortgage (servicing the debt), to ensure that the building will generate enough revenue to cover the mortgage.

At the very minimum, the NOI must be equal to what the borrower has to pay in debt service. In other words, the ratio must be at least 1.0 in order for the property to break even on mortgage payments. However, this ratio, where total income exactly equals the cost of servicing the debt, allows no margin of error should economic circumstances deteriorate; for instance, if a tenant vacates or a recession hits and rents go down.

Most lenders require a DSCR of 1.2 and up to provide a cushion in case the NOI goes down for any reason. This means the building should produce 20 percent more income than the borrower needs in order to repay the debt. The DSCR calculation is NOI divided by debt service.

$$DSCR = \frac{Net\ operating\ income}{Debt\ service}$$

For example, if the cost of servicing the loan is $100,000, and the NOI is $120,000, this would result in a DSCR of 1.2.

$$\frac{\$120,000}{\$100,000} = 1.2$$

In this example, even if the NOI fell $20,000, the borrower could still service the debt. In such a case, however, any refinance would need to be at a lower level of debt in order to restore it to an acceptable DSCR. If NOI dropped down to $100,000, it would require a recalibration of how much the bank could lend in order to reduce the debt service level so that the $100,000 of NOI also fits within the 1.2 ratio for the DSCR—which, in this case, would be $83,333 of debt service against $100,000 of NOI.

If the DSCR falls below a 1.0 ratio, it costs the owners money. This is because the property's income is less than the debt service payments. Here, the owners have to dig into their own pockets every month, collect the rents, and use other resources to pay the mortgage. Consequently, since this significantly increases the default risk—which is the reason why most lenders require the 1.2 ratio or greater—neither borrowers nor investors should be comfortable with a 1.0 ratio.

In addition, DSCR levels vary according to the asset type and its associated perceived risk. For example, multifamily residential deals typically require a DSCR of 1.2 or higher. For commercial assets which have fewer tenants contributing a higher proportion of the total rent—a perceived higher vacancy risk—lenders may require a 1.5 DSCR in order to extend a loan for the property.

In rare circumstances, the DSCR can be negative. It's unusual and typically found only in value add deals where the borrower actually provides an interest reserve. If there is a negative DSCR—no rents, or very limited rents—during the value add construction period, a lender may compensate for that by requiring the borrower to contribute, up front, one to two years of debt service before they agree to make the loan.

DSCR is a good method for testing a deal's vulnerability during an economic downturn. When considering a deal's DSCR, a high ratio should give you confidence that in the event of a downturn, the sponsor will have enough wiggle room to ride out the storm. By definition, low-leverage deals have much higher DSCRs because these represent a higher tolerance for decreased NOIs during a recession or economic downturn, and so greater resilience to recession.

Again, this speaks to your risk-return profile. If your appetite for risk is higher, then a low DSCR won't bother you. If you have a longer-term view and a lower appetite for risk, then you want a good cushion with the DSCR.

To wrap up this concept, let's take a look at using DSCR as a risk-reward indicator. Here we have two buildings, one in Deal A and one in Deal B, which are exactly the same in all regards, except for what is shown in Figure 8.10.

Deal A has a DSCR of 1.6 with a pro forma IRR of 14 percent. With a high DSCR, and a 14 percent IRR, this deal is perhaps recession-resilient, having a higher tolerance to a drop in NOI. If the NOI comes down, the borrower can continue making the mortgage payment. However, with this lower risk profile, there may be a lower return. It's a more conservative play.

In Deal B, there is a DSCR of 1.2 and a pro forma IRR of 18 percent. If a sponsor comes to you with this type of a deal, they have maxed out on their finance. Compared to Deal A, Deal B carries a higher risk of default: If the sponsor's projections are wrong or the NOI decreases for some reason, it will be harder to make the mortgage payments.

Deal A

- DSCR 1.6
- Pro-forma IRR of 14%

- Recession resilient
- Lower risk, lower return
- More conservative

Deal B

- DSCR 1.2
- Pro-forma IRR of 18%

- Maxed out finance
- Higher risk of default
- More susceptible to recessionary pressures
- Higher return

Figure 8.10 Deal A and Deal B

Deal B is more susceptible to recessionary pressures. However, in exchange for the higher-level risk, it offers a higher return.

Notes

1 Cash-on-cash returns are the actual distributions made relative to the amount of money invested at any given point in time, though most frequently based on annual pre-tax cash flow. They are sometimes referred to as the 'yield on investment.'
2 You can hear the entire conversation with Dr. MacKinnon here: https://gowercrowd.com/podcast/018-real-estate-value-investing
3 Larry Feldman of Feldman Equities discusses the comparison between how the cap rate is used in real estate and how the concept of the 'multiple' is used in valuing stocks or other equities. You can learn more about his insights in my conversation with him, available here: https://gowercrowd.com/podcast/larry-feldman-ceo-feldman-equities
4 Some sponsors may prescreen deals on the basis of having a certain spread between market and build to cap rates, though they may say 200 basis points, or 200 bips, or 200 bps, or 2 percent spread—all refer to the difference between the cost and the market value.
5 As wacky as this sounds, to get 0 percent loans is not entirely without precedent—though the only example I've ever seen was not for mere mortals like thee and me. During the global financial crisis of 2007–2008, the FDIC shut down banks that had excessive nonperforming loans on their books, mostly collateralized by real estate. They pooled these loans into vast portfolios and then auctioned them off to institutional investors. The idea was that the taxpayer should not foot the bill for cleaning up these failed banks, so the auctions were not straight sales; rather, they were invitations to invest *with* the FDIC in partnerships. In the deals I worked on and saw, the FDIC sold 50 percent shares in newly formed entities that owned these huge portfolios in the hundreds of millions of dollars, and they offered 0 percent loans to buyers to incentivize participation.
6 I have written extensively on this topic and the impact it also has on the co-invest that sponsors are expected to contribute to deals. You can find out more about this, including to some video presentations I've put together, here: https://gowercrowd.com/learn/impact-of-crowdfunding-on-real-estate-waterfall-structures
7 You can learn a great deal about a sponsor by the way they discuss this topic. Most real estate crowdfunding sites have good educational content about the IRR and other topics, but far fewer real estate sponsors do. When you're looking at a deal, be sure to refer to the sponsor's website for educational articles that explain what their take is on the IRR, its pros and cons and how they use it in their underwriting and projections. And if they don't have anything in writing already on their site, be sure to ask them before making your decision.

8 See my white paper on waterfall structures for a deeper dive into this concept at https://gowercrowd.com/learn/whitepaper

9 The exceptions to this rule that I've seen have only been technical. I saw a handful of deals during the downturn of 2007–2008 where the bank expected to be paid first, but it was discovered that the proper paperwork had not been filed when they first made the loan. When it came to deciding who had rights to the proceeds of the sale of the property and evidence was presented during litigation, although it was obvious that the bank had lent money, they were unable to present evidence for exactly when they had lent it. In being unable to prove their position in the hierarchy of the waterfall, they lost their rights and somebody else took priority for getting paid.

10 See the section in this chapter on promote and fees.

11 Be very wary of anybody that says your investment is guaranteed. When I was first in real estate during the 1980s, it was a lesson I learned from watching the Savings and Loan debacle unfold. There were characters at the heart of that crisis who had 'guaranteed' investments and who ended up spending time at Club Fed when their investors eventually lost all. I am sure that there are some notable exceptions that are worth taking a closer look at, but I can only think of one: Steve Kaufman at Zeus Crowdfunding in Houston guarantees his investors' principal. I was skeptical when I first heard that he did this but when I spoke to him, he made a compelling case for how he justifies the guarantee. You can listen to the conversation I had with Steve in series 2 of my podcast and decide for yourself. It's available at: https://gowercrowd.com/podcast/235-steven-kaufman-zeus-crowdfunding

12 See the section in this chapter on leverage.

13 The most common preferred return paid to investors in real estate syndications at time of writing is 8 percent, used by 40 percent of deals; the next most common pref is 10 percent used by 30 percent of deals; 6 percent and 12 percent are the next most common; and the remainder vary between 0 percent and 22 percent.

14 One of the main reasons that institutional deals are structured very differently from deals made by private individuals, even if they are of similar scale, is that in the institutional world information about deals is much more efficiently distributed. There are relatively few institutional players. They are, by and large, seasoned, very experienced professionals who all know each other and depend on personal reputation to transact. The availability of information and access to resources to double-check sponsor due diligence is far more advanced than in private deals, and the terms and conditions and rights and responsibilities of all parties are negotiated by extremely expensive, highly experienced attorneys, all fighting for their clients' best interests. The availability of knowledge is equally shared between investor and sponsor in the institutional world, whereas private individuals do not enjoy these conditions. No matter what anyone says to you, in most cases you are not investing 'like an institution.'

15 See the section in this chapter on promote and fees.

16 See Addendum 1 for a more detailed examination of preferred equity, in an article that first appeared on gowercrowd.com

17 See the section in this chapter on promote and fees.

18 You can get access to the white paper at: https://gowercrowd.com/learn/whitepaper There is also a series of videos on my website in which I walk you through the research we conducted and discuss the findings in detail.

19 See the section on IRRs in this chapter.

20 I have included one of the articles I wrote about waterfalls—"The impact of crowdfunding on real estate waterfall structures"—in Addendum 2. In it, I share some very important concepts that you can use to evaluate waterfalls and sponsor integrity.

21 See Addendum 1.

22 Keep in mind that with preferred equity positions in second place behind the primary lender, the process for taking control of a project is significantly easier than the foreclosure process. See Addendum 1 for more details on this.

23 See Chapter 9 on contracts for a discussion on the dilution issues and what to watch for.

24 At the Iowa State Fair in August of 2011, candidate for president of the United States Mitt Romney is quoted as saying "corporations are people, my friend," in response to a heckler's cries from the audience.

9 Contracts[1]

When you see a project online that you like and are thinking about investing in, there will come a point where you are going to be invited to invest. In the digital world this decision point will be represented by a button on your screen that you can click on to "Invest Now!" Behind those buttons, the contracts that you're expected to sign (digitally) to be accepted into these deals are extraordinarily long. In fact, they can be 80,000 to 100,000 words long. That's about as long as this book.

The contracts may come in the form of one document or a series of documents that are collectively called the 'offering documents.' In this chapter, we'll review how to read through the offering documents without being overwhelmed by their length, complexity, and legalese.

There are four key issues to look for, which should drive your decision to invest or not in the deal. When you get to that 'invest now' moment and want to read the documents, these issues will help you determine whether or not a deal is suitable for you and will help you recognize when the sponsor is being fair and reasonable. In other words, by understanding the key issues, you will be able to work out if you are getting a good deal from the sponsor for the money that you're investing.

In this chapter, I'll share insights that I've gained from experiences from my own career together with those gleaned from conversations with some of the top real estate attorneys in the United States.

Learning from experience (mine and others')

Throughout my 30 plus years of experience in real estate investment and finance, I've been through multiple cycles. I've lived through two major real estate downturns and, during those downturns, I've been exposed to many contracts and scenarios, both contracts that I've signed and those that I've seen others sign.[2] After sharing insights from these various perspectives, you'll understand:

1. what happens when deals go wrong;
2. what you need to look for inside these contracts;
3. how to protect yourself.

Keep in mind that contracts are not written for when things go right; they're written for when things go wrong.

The effects of a downturn

The first downturn I experienced was in the late 1980s and early 1990s. It was precipitated by the savings and loan crisis. The second was the financial crisis of 2007–2008 during which there was a near systemic failure of the economy. Having been through those meltdowns, I've learned to read between the lines of what sponsors actually write in contracts. It is tempting to take the language in contracts at face value; however, without knowing the background as to why something is in a contract, you can miss important nuances that are buried in the legalese.

Becoming a more conservative investor

One side effect of experiencing two major downturns, particularly the most recent one where so many deals went bad, is that I have become an extremely conservative investor. During the first downturn in the late 1980s, I had signed investment contracts to be a member of an LLC to invest in deals. I admit that I didn't have the slightest clue what I was signing. To this day, I remember running through the contracts and going straight to the section that explained how much money I would make.

At the time, I was in my early to mid twenties. I spent all the money that I earned in salary—I bought a brand-new Mercedes, lived in a fancy apartment, and enjoyed some high living. In my mind, I was making millions of dollars through these real estate deals that I was involved in.

However, what became known as the savings and loan crisis precipitated the failure of thousands of banks nationwide. When the market collapsed, I lost absolutely everything. But since I'd also taken on debt—I'd actually borrowed against what it was that I thought I was making—I ended up losing more than I had actually invested. I had borrowed against the money I thought I had made. I was a beginner, made some bad mistakes, and I learned how not to make the same mistakes again.

Before the Great Recession of 2007–2008 happened, I had had the foresight, in 2007, to sell my entire real estate portfolio. I entered the downturn with no legacy problems and with a lot of dry powder that I could invest in real estate deals. Although I spent a lot of time looking for opportunities, I couldn't find any. There were a lot of problems in the real estate industry and everyone had hunkered down. To say it was a real disaster is no exaggeration.

In search of bargain-basement real estate deals, I made hundreds of millions of dollars in offers to banks to buy properties directly from them by acquiring the loans that I knew were under water—meaning that the value of the real estate that collateralized the loans was worth less than the face value of the loan itself. I would send faxes (yes, faxes) and emails and make calls, but I received no responses from anyone at any bank. I came to understand later that the reason for this silence was that most banks were suffering from existential loan default crises and had no idea how to resolve their problems. Bank employees were torn between the bonds they had established with borrowers with whom they, in some cases, had close personal friendships and foreclosing on those borrowers and pursuing legal action against guarantors.

One of my prior investors was on the board of directors at a major California regional bank that had done a lot of real estate collateralized lending. He introduced me to the chairman of the bank and I was hired, as a non-banker, to help clean up their balance sheet. As I looked at the bank's entire portfolio, I saw hundreds, if not thousands, of borrowers—real estate investors and real estate developers—whose borrowings were significantly higher than the value of the real estate they had borrowed against.

By examining the bank's multibillion-dollar portfolio, I saw almost every single way for deals to go wrong. It was an enlightening period for me. I gained an appreciation for understanding what's in the contracts. Furthermore, it showed how much power a senior secured lender has in any deal, and I began to deepen my understanding of the influence a bank has in any deal.

Of course, as the overall economy pulled out of the slump, the real estate market started to improve dramatically. On top of that, the passing of the 2012 JOBS Act created an exciting new way to invest in real estate deals online. But under the surface, it also created a potentially risky way of investing in real estate unless you really know what you're doing.

In this chapter, I'll show you how to navigate the potential minefield of online deals through the lens of these contracts so you can identify the true winners worthy of your investment. Specifically, we'll look at:

- how to look at contracts' DNA so you can invest with more confidence and be sure you're in a deal that makes sense to you;
- how to avoid deals which don't offer the kinds of rights you intuitively expect to have if you're sending money to someone;
- how to get through enormously long contracts without being overwhelmed by their complexity;
- how to take shortcuts to find the key sections that drive your decisions;
- how to stop wasting time on bad deals and move on quickly to the next one until you find a deal that is suitable for you; and
- how to drill down to greater detail and decide whether or not to wire your money.

Building wealth through passive income

It's easy (and I say this tongue in cheek) to make money and build wealth in real estate by investing with somebody else. After all, you let them do all the work that you would otherwise have to do. But in reality, it isn't a cakewalk. When investing in real estate, there are two basic concepts to discuss: wealth preservation and growth; and ongoing (passive) income streams.

Both of these are highly desirable. After all, everyone wants to build wealth. And we all want passive income. But not all 'passive income' is created equally, and while the idea of earning income without having to be active might be attractive, the way it is used outside of its original context (as a definition within the tax code) has become oversimplified and distorted.

Passive income

Apart from the Internal Revenue Service definition of 'passive income,' which I will NOT be discussing here,[3] there are two types of passive income to be aware of. The first type suggests the application of minimal effort to receive income: We make an investment and money automatically comes in. We don't technically work to see those checks arrive. Usually, we have a day job but we also have investments which supplement our daily income; our weekly envelope. That is one type of passive income.

Yet it is important to distinguish between that type and passive income as an investment where you are passive in a deal. Translation: You have no rights in that deal. Maybe things go south, or maybe the sponsor does things that you didn't expect and that you don't like. But you can't do anything. That type of passivity in a real estate investment is something you really don't want, or maybe you don't care about that. Regardless, we'll see how to recognize if this is the kind of passive investing you're getting into so that you can make an educated decision about how much true *passivity* you are willing to tolerate.

While the concepts in this chapter are for beginners, they're also applicable to sophisticated investors, accredited investors, and non-accredited investors. I'm introducing beginner-level concepts because I was once a beginner, and without knowing these lessons, I lost everything. If I had known to look at contracts the way I look at them now, I could have gone straight to being a more sophisticated investor, bypassing the beginner phase completely—as you are doing now by reading this book.

Throughout my career, I have reviewed many thousands of deals, not only as a principal, but also for major institutions—for the bank and for one of the world's largest private equity funds. As I mentioned above, during the first downturn of the late 1980s, I lost everything that—in my mind, at least—I had already made.

Since I never want that to happen again, I've become super-conservative, almost to the point that the only deals that I personally invest in are long-term deals. In general, I prefer to have a long-term perspective and the ability to bring enough money to a deal to influence the terms. So using my own experiences, I'll show you how to look at contracts and how to see them from the same perspective.

Power and the capital stack

At the highest level, the first thing to understand is how power and influence in a deal flows through the contracts. First, let's contextualize one of the most important concepts relating to how these contracts are structured (which is discussed at length in Chapter 8 in the section on leverage): the capital stack.

The capital stack is the industry term used to refer to a deal's financial structure. Specifically, the way money comes into the deal is like a stack of money. At the bottom of the stack is the lender and the debt, representing the foundation of the deal. Here, at the foundation, is where the bank retains the greatest power in any deal. Everything is built on top of that foundation of debt. Above that are various different strata of equity, including preferred equity or common equity. There could also be a mezzanine debt layer which bridges the gap between equity and debt.

In industry speak, each one of these layers is referred to in terms of how it gets paid off; as money comes back into the deal, the bottom of the stack (the debt) is paid off first. Each layer gets paid off in sequence as you move up the stack, with the equity layers—your money—getting paid off last. The terminology implies that you get 'paid off.' It's a positive feel-good language pattern that can lull the unaware into a false sense of security.

The capital stack is a balancing act of power. Whoever sits at the very bottom of the stack—typically the lender—has the power over everybody else in the capital stack. The lender has power to foreclose on and eliminate any hope that anyone else in the stack will recover any of their money. Also, each layer of the capital stack has that same power and influence over the layers above. This is the significance of the capital stack. And it's that theme of power (amongst others) that cascades through every contract you will look at.

Learning about the capital stack from within a bank

During the second major downturn that I lived through, I began to see the power of the capital stack and how it worked.

In 2007, I saw that something was wrong with the industry and the economy and I sold my entire real estate portfolio. Going into 2008, I saw the world through the vantage point of working for a bank. I learned and realized the strength and authority that a bank—or any senior secured lender—has over everybody else in the deal. As the priority creditor in a deal (i.e. the one 'paid out first'), they can wipe everybody else out. In some cases, it isn't easy for the lender to do that, but ultimately the odds are stacked in their favor. That bank-insider experience gave me an appreciation for the power the right to foreclose on other layers of the capital stack provides.

Think about the contracts that define the different layers in the capital stack as the mortar that binds the layers to each other. The agreements that each layer has with the other layers define the strength with which each of these layers is bound together and the way in which they influence each other.[4]

The first thing to understand when looking at the contracts is that signing a contract, even digitally, is a commitment to the deal. When you hit the 'I Agree' or 'Invest Now' button, you make a legally binding commitment to invest in the deal. Figure 9.1 shows a screenshot of a typical 'invest now' set of offering documents.

This particular example is big—it's about as long as this book—and one of the reasons why they're so huge is that the securities laws mandate that investors have enough information to make informed decisions. So, in theory, this vast array of legal mumbo jumbo is designed to protect the investor.

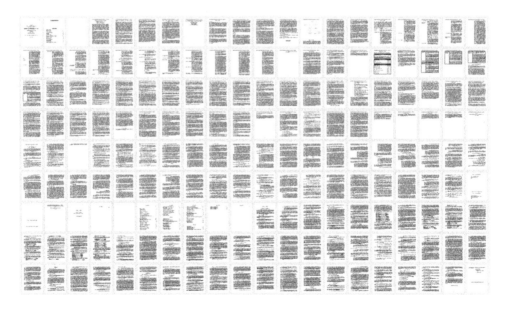

Figure 9.1 Typical 'invest now' contract

Four key themes

Since contracts bind the layers of the capital stack together, it is important to understand that they are written this way for two primary reasons. Firstly, the contracts dictate terms for when everything goes according to plan. In this scenario, contracts simply explain everyone's rights and responsibilities and how you get paid out, how much you get paid, and when. Secondly, and more importantly, the bulk of a contract covers what happens when things go badly. As an investor, knowing what to expect if things go south is extremely important.

In every contract, there are four consistent themes running throughout. By understanding these four key themes, you'll be better equipped to distinguish a suitable deal from one that is not suitable for you.

1. The golden thread that runs through every clause in every contract is the idea that real estate is a **zero-sum game**. By grabbing the golden thread between the sponsor and the investor, either side can draw more to their side and give less to the other side.
2. The second key theme is the idea of **investor rights and remedies**.
3. The third key theme is **sponsor responsibilities**.
4. The fourth key theme is the idea of **power and control**.

All four of these themes run through all the contracts and all the clauses. And keeping them in mind will help you to contextualize what each clause in any contract means.

Three important questions

The following three questions will help you consider these four concepts.

1. What is the risk to investors?
2. What's the sponsor's risk?
3. If things go wrong, what rights and remedies do you have?

What's your risk?

First and foremost, before committing to a deal, double-check that your investment, no matter how much it is, is all you are committing to and nothing more.

A quick way to search for relevant information within incredibly long contracts is through keyword searches. Be aware that this isn't a foolproof method, but it is a great way to get to the key terms very quickly. For example, I ran the term 'capital' through the 80,000-word document shown in Figure 9.1, and I got 245 hits.

Since 245 is still a lengthy list, I narrowed it down with a keyword search on the term 'capital call.' This is the term typically used in contracts to describe the right a sponsor retains to demand committed investment capital from investors. With that term, I got five hits. One of those hits is a clause in the contract which says:

No additional capital contributions shall be required of the members, in addition to the initial capital contribution of the members in the amount set forth as the purchase price in each member's related subscription agreement.

In context, the phrase "no additional capital contributions" is fairly clear. In this particular example, there is no additional requirement for money. That is fairly common, though still something you want to confirm.

The timing of when your capital may be called is also something to consider carefully. What does the contract say about timing? When do you contribute that money? If you are expected to contribute the amount all at once and you agree, you sign and wire money. But what happens if what you're agreeing to commit now and wire the money later? What happens if you agree to wire part of the money now and more in six months? This seems reasonable, maybe even attractive, but what happens if the deal goes wrong between the time that you commit and when you must send the remainder of the money? You could be liable for wiring money into a deal that has already gone south, because you inadvertently committed to doing that. This is a very important thing to look for and to avoid.

How you'll get paid out

Another key thing to look for is how you'll get paid out. To explain this, you need to understand exactly what you're getting for your money. To help, Figure 9.2 illustrates the basic structure of a deal.

In real estate crowdfunding, in most cases you are not buying real state; rather, you're becoming a shareholder in a company. The developer will form an independent company that is a brand-new entity formed specifically for the project that purchases the real estate. That company will then hire the developer to provide services to run the company and to develop the real estate. As an investor, you are actually investing in the company, buying stock in the company in a similar way as you buy listed stocks (except you cannot trade these real estate stocks), which is why the contracts are governed by securities laws. As a result, you become a shareholder who gets returns for the money that you invest.

To reiterate, the contracts are the mortar, the binding agent between the company and you. It is the contract between you and the company that defines how much you will make. So if you buy a 10 percent share of that company's stock, that does not necessarily mean that you own 10 percent of the profits—unless the contracts specifically say that you do. In fact, somebody with no ownership of the company might own 100 percent of the profits of the company, and vice versa; somebody who owns 100 percent of the company might actually own 0 percent of the profits.

Figure 9.2 Typical deal structure

The variation between those two polarities is entirely at the discretion of the sponsor in conjunction with the attorney who wrote the contracts. They decide exactly how those different relationships are structured. There is no standardization across contracts in the industry. Every single contract is different. There could be absolutely any structure of any kind to define what rights participants have, who gets what, who gets to vote, and other issues; the range of possibilities is endless.

Generally, the different ownership types are limited only by the sponsor's imagination. There's common stock, preferred equity, convertible preferred equity, convertible notes, and many more. Yet it is the sponsor who defines each relationship according to the offering documents.

Deal structure and getting paid out

How you get paid out also depends entirely on the contract structure and how it is written. You must look for consistency in the contracts between the pitch deck (the glossy, sizzly sales materials) and what's actually in the offering agreements. By the way, the more complicated the deal structure, the harder it is to calculate how you get paid. Let me emphasize that point: Who calculates the exact amount that you get paid? Do you calculate it and send an invoice? No! You sit passively and wait until the sponsor wires you money. But who calculates that? What calculations do they use to determine exactly how much to send you?

I ran the keyword 'distributions' across that 80,000-word contract in Figure 9.1 and got 23 hits. 'Distributions' is the word most commonly used to describe how investors get paid. Here's one of the hits that I got—this is typical of the kind of structure that you see when you invest.

> *And in the event that investors have achieved an internal rate of return equal to eight percent, any and all amounts that would be distributed to the investor LLC, first 20 percent to the managing member and 80 percent to the investors, until the investors achieve an internal rate of return equal to 14 percent. Second, 35 percent to the managing member and 65 percent to the investors, until the investors achieve an internal rate of return equal to 20 percent.*

How do you calculate that? In order to calculate the IRR, you need a terminal value. This means that until the sponsor sells or values the property, it's impossible to calculate those numbers. If the sponsor doesn't sell the property but you receive dividends, like income from rents or other distributions without a sale, how is the building valued to calculate those IRRs? Where do those hurdles come in, and who is deciding how to value the property?

I'd bet you a pound to a penny that the sponsor—the developer—has all the rights to define that or to determine value. And of course, the sponsor also calculates the returns. Without a doubt, you want to understand how the sponsor makes their calculations so you can be sure they are doing it accurately.

What's the sponsor's risk?

The second question covers sponsor responsibilities, how much the sponsor is risking, and how they make money. Overall, there is one guiding principle to understand whether or not a deal is suitable for you: the alignment of interests between you and the sponsor (or lack thereof).

Key things to check that there is an alignment of interest include:

- whether the sponsor is investing in the deal themselves with their own money;
- what their fees are;
- whether they are taking enough fees up front to contribute their share of the investment into the deal; and
- whether the sponsor includes added value to make their co-invest in some way that does not require them to bring fresh money to the table.

In short: Are they are taking the same risks as you? Are their interests aligned with yours?

What rights and remedies do you have?

The third key question to ask is this: In the event that things go wrong, what are your rights and remedies, and who has power and control in the deal? In most syndicated real estate deals, you may have neither power nor control. Period. The sponsor controls everything. Indeed, some investors will invest through investment groups or clubs. These groups and clubs typically have their own offering documents that in many cases can add distance between the investor and the sponsor and further barriers to control or influence and to investor rights and remedies. In that case, it becomes increasingly difficult to protect your own interests. That's not to say you shouldn't invest through a club or private syndicate, but be sure to understand who retains the control over those operating entities, as it will be them you will be relying on should anything go wrong.

Returning to the idea of passive income, here's the reality: Being a passive investor in these deals means that if things go wrong, you likely have no control over how the situation gets resolved. You cannot protect your own interests at all. You are totally, completely passive, maybe even without voting rights. You might not have the right to know who the other investors are. Think about that for a moment. What happens if you want to call a vote, but you don't know who to go to? A very tangible example of why that is so fundamentally important relates to the idea of dilution. Remarkably, you might not even have the right to know that you're being diluted in the contract.

This directly contrasts with, for example, the way that institutional investors negotiate to preserve their rights. Those rights might include things like veto rights over major decisions, such as the sponsor's ability to take on more debt or if and when the sponsor can sell the property. An institutional investor may also maintain the right to veto, allowing new investors into the deal. In addition, they may retain budgetary controls—a constant monitoring of what is going on—so that they can step in if things go wrong. In all likelihood, institutional investors may also maintain the right to remove the manager and take over.

I saw this happen dozens of times during the last downturn. Private equity funds (institutional investors) realized that things were going south and they stepped in. In the end, they took projects away from developers. Yet this power to step in depends on the contract. After a quick search in the 80,000-word contract for the word 'remove,' I got seven hits. One of these said:

> *The members, upon vote or written consent of a 75 percent majority may remove the manager, but only for cause. Cause shall mean a final non-appealable determination by a court of competent jurisdiction that the manager has committed fraud.*

According to this contract, the only way to step in and remove the sponsor is if they commit fraud. First of all, that's a huge legal hurdle and very difficult, time-consuming and costly to prove. Secondly, that would only happen after going through a court case to get them out. Even if sponsors are incompetent, careless, lazy, take too many vacations, never answer your phone calls, or never respond to your emails, you cannot do anything about it unless you have some basic rights.

Doom and gloom or silver lining?

Although some of this information seems like doom and gloom, there is a bright side. Real estate is a phenomenally exciting way to invest, and you can make a lot of money doing it. Through real estate investment, you can build wealth and earn passive income without actually being passive in the way that I've described above. In the end, real estate investing has great potential to be a lucrative and exciting way of supplementing your income over time.

Being fully informed is the true key to successful investment. It's actually what the securities laws demand. But you have to read between the lines. In many cases, you need the benefit of hindsight in order to understand exactly what you're getting into. In the next section, I'm going to walk you through some of the key things to look for in investment contracts as well as provide you some tools you can use to protect yourself.

Contracts: Deep dive

When you look at a deal, you want to see it clearly. This section provides a global perspective, a high-level overview of how to look at these really huge contracts. Then we'll review some specifics you can look out for as you assess the offering documents on any deal.

Three big ideas

First of all, let's explore three concepts that encompass everything. When you look at a contract, always keep these three ideas in mind. The first one is the concept of debt—what it means and the specific implications of debt. Second is the concept of risk aversion. When you look at contracts, keep in mind that everybody involved is risk averse. Third is the concept that contracts are inherently adversarial, because, as I mentioned earlier, real estate is a zero-sum game. Whenever two parties go into an agreement together, the contract is inherently adversarial, as both sides tussle to get as much as they can.

Debt: Layers of power and influence

Debt changes everything. What I mean by this is that when you put debt on a real estate deal, what you're really doing is creating layers of power and influence, and most of the power and influence in the deal vests with the senior secured lender.

What really changes everything is that real estate is highly leveraged. There are layers of debt. The leverage creates incredibly 'juiced' returns. For example, in most online deals, you see returns at the low end, 8, 9, maybe 10 percent, with 15, 20, 25 percent IRRs projected at the high end. Such high returns are only possible when there is debt levered on the property.

Yet, when a lender lends to that property, it gives them supreme power over everybody else in the capital stack. Their priority in the stack gives them the right to foreclose on everyone, and if they need to, they will undoubtedly exercise that right. They will do this because they have a responsibility to their shareholders and because regulations mandate that they do.

More importantly, federal law heavily regulates lenders. If things start to go south, the lenders will dominate any kind of negotiation because they have zero margin for error. Furthermore, they have considerable influence over the court process. Debt is a very important component in any deal for various reasons, and it really doesn't matter where you are on the capital stack, the lender is always ahead of you.

On the other hand, if there was no debt on a property, the likelihood of a default goes down to almost zero. When you think about it, without debt or borrowing on a property, there can be no default on a loan, because there isn't a loan.

In this context, you can think about contracts as being largely driven by the 'what if' question: What if the lender calls the loan? Overall, it is worth looking in contracts for defenses against any kind of recovery provisions that a lender may have and asking how you can protect your own interests in the event of a default on the loan.

Risk aversion: Most people have it

The second way of conceptualizing how these contracts work is through understanding that people are risk averse. The pain of loss is typically much greater than the joy of gain, but if the perception of potential loss is skewed, then you may not make some decisions that you would otherwise make.

Let's say that you're looking at these contracts and you think that the possibility of losing your money or of there being a problem is actually much less than is, in fact, the case. You might make the wrong decision, which puts you in a situation that you don't want to be in—one where you could suffer a loss.

This is what both sides are trying to defend against in contracts. The sponsor tries to insulate himself against the possibility of loss. At the same time, investors try to insulate themselves against the risk of loss. Everybody knows that they're getting into business to make money. That's easy. That's the only reason anyone's sitting down.

However, a contract's structure is designed to offer protections against loss for both sides. But because it is a zero-sum game, when you're defending against loss, you may also be pulling against each other's interests in the contract. In every contract, you see this idea repeated time and time again.

Let's look at an example that contrasts avoidance of pain through suffering loss with delight in making money. In Scenario 1, there are two options:

- Option 1—a 100 percent chance of getting $3,000; or
- Option 2—a 75 percent chance of getting $4,000 (a 25 percent chance of getting zero).

Which option would you choose?

In Scenario 2, there are also two options:

- Option 1—a 100 percent chance of losing $3,000; or
- Option 2—a 75 percent chance of losing $4,000.

Studies have shown that in the first scenario, most people avoid the risk and take the first option; the certainty of a $3,000 gain. On the other hand, when presented with the second scenario, most favor the second option—a 75 percent chance of losing $4,000, because it offers the possibility of avoiding pain of loss.

As you review the contracts, keep in mind that both parties are fighting against each other to avoid the pain of loss. If somebody loses, then it means that somebody else potentially gains. You always have that dichotomy and that tension is a common thread running through the contracts.

Contracts: Inherently adversarial

The third theme that runs through these contracts, and through most of the clauses, is linked to the first two themes. You'll start to see, as you go through contracts, that they are inherently adversarial. Developers write them to defend themselves against potential loss and to maximize their own gain. They want to control as much as possible. Yet, investors' interests are very similar: they too want to defend against loss and to maximize gain.

In any kind of business agreement, there is typically a first-draft advantage to the person who puts out the contract. Even though it costs more money to draft and present the contract the first time, it's generally considered advantageous to be the one who authors the initial contract, because it is a defining foundation upon which all future negotiations will be conducted.

In 'take it or leave it' contracts, like you will see in real estate crowdfunding, that's an absolute advantage to the sponsor. These are no-negotiation contracts, so your only option if you don't like the contract is to walk away. If you want into the deal, you take the terms. In reality, a 'take it or leave it' contract isn't a first draft—it's a final agreement, and the contract writer already has the advantage.

In a 'take it or leave it' contract where there is an inherently adversarial relationship, make sure that you are not too far on the losing end. When you look at the terms, determine whether or not you can handle them going forward.

Below we are going to look at some key rules and concepts to keep an eye out for. First of all, let me point out that these are very risk-profile specific. Some may be specific to your personal circumstances and your own personal risk profile and so may resonate with you. But it is difficult to give greater weight to any in particular.

Four key rules

1. No matter what you invest in, understand that you could lose everything. Don't invest anything you can't afford to lose, and don't invest anything that you might need before there's a likelihood of it being returned to you.
2. Keep in mind that the senior secured lender, the bank, has ultimate power over everything. Their rights are superior to those of everybody else. And that applies to ALL levels of the capital stack. Even though anyone ahead of you in the capital stack has control over you, nobody has as much control or power as the senior secured lender.
3. The further away you are from the source of the deal, the harder it is for you to recover losses or to control what's happening. The developer has the most control. The investor has the next level of control. However, if you invest through an

intermediary like a group or a club, you are now an extra layer away from control. You want to know how difficult it would be to pierce the veil of multilayered protections between you and where the money is actually being deployed.

4. Finally, the biggest takeaway is this: If you don't understand the contract or some of the key sections, don't invest in the deal. Honestly, investing in a contract you don't understand is almost as bad as not reading it at all. **Do not** read a contract and then, without understanding it, sign it and wire your money. Make sure you understand it. If you don't understand or if you can't explain it in simple terms to your better half or to somebody who simply doesn't get real estate, the deal is probably too complicated. Don't invest in the deal.

Ten key things to look for in real estate contracts

In this section are ten key things to look for. Again, these may not fit within your own personal appetite for risk. Maybe you can accept them, or maybe you can't. Regardless, these are ten keys to be aware of.

1. Rights and remedies
    The first thing to look for are your rights and remedies. Although you never want to fight a lawsuit, here are some important questions and observations regarding rights and remedies.

    *   You should know the jurisdiction in case there is a lawsuit.
    *   Ensure you have rights to remove the manager under certain circumstances.
    *   What kind of rights are you going to be comfortable with?
    *   How does the voting look within the contract?
    *   Do you have the opportunity to vote with the other members?
    *   What are the circumstances under which you can take a vote, and what is the mechanism for calling a vote and for actually having a vote?

2. Dilution issues
    The second key thing to look at is dilution. In particular, what happens if there is insufficient capital? Maybe the borrower runs out of money, goes over budget, or didn't raise enough capital. Maybe they got their numbers wrong somehow. Can the sponsor bring in investors ahead of you, ahead of the seed investors, and dilute them? More specifically, can they dilute *you*? Also, what are the circumstances by which a capital call can be made? Can the sponsor borrow from a bank to do the deal instead? If shortfalls exist, will the sponsor go to the members for a loan, or will there be a dilutive capital call of some sort?
    In a well-reasoned contract, before a sponsor takes on more debt, he should at least ask the members (investors) for a loan with reasonable terms.

3. What happens if the loan is called
    Another key thing to look for is what happens if the loan goes into default and is called by the lender. Will you need to continue ponying up more money in a bad deal? For example, maybe you've agreed to stagger your contributions to the project. But what happens if the deal goes bad and you still haven't made all contributions that you committed to earlier? Are you going to fund a distressed asset? In other words, will you be a de facto guarantor in some way? Be sure to look out for wording that gives that impression.

4. The guarantor

    The fourth thing to look out for pertains to the guarantor. In California, if the guarantor and the entity developing are one and the same, then the guarantee is unenforceable; it is a sham guarantee. So, keep an eye out for that. Also, since lenders (or whoever's in first position) take everything, you may want a real guarantor between you and the secured lenders. This might mean having an unrelated third-party guarantor who guarantees the loan. That way, the likelihood that the lender will call a default or foreclose will be mitigated by having a real guarantor in between you and that lender, protecting the downside.

5. A continuity plan

    The fifth key thing to look for in these deals is a continuity plan. Look at whether the deal has a single-member manager and whether there is only one person operating the deal. If you see that there is only one person, you may want to be cautious about investing, because, if something happens to the person who is in charge, who's going to be around to manage your investment? You also want to be sure that there will be a backup plan, also called a continuity plan.

6. The sponsor's own risk profile

    The sixth key thing to look for is the sponsor's own risk profile. Does the sponsor make money even if you don't? If you lose money, does the sponsor lose money? In other words, is there an alignment of interests between you and the sponsor? What fees do they have, and how do these relate to their co-invest money? Learn whether their fees are front-loaded. For example, if they take a half-million-dollar deal fee at the beginning but their contribution—their co-invest—is the same or less, do they really have skin in the game?

7. A liquidity provision

    The seventh key thing to look for is a liquidity provision. Imagine a situation in which you and your heirs need your money back for a legitimate emergency (not because you want to go on a vacation). If something happens, you want to be sure that there is a contractual provision that defines how that will be managed and how your position can be liquidated.

8. The waterfall

    The eighth key thing to look for is the waterfall, the way proceeds are distributed to investors. In particular, pay attention to complicated waterfalls, especially with IRR metrics that are central to the way people are paid out. Make sure that the sponsor has a specific plan in place for calculating those returns when the time comes. And make sure that all the documents are consistent. Specifically, verify that the pitch deck and the offering documents say the same thing.

    It is also important to make sure that the disbursement method isn't driven by the tax code. Be sure that there is a clear explanation, not wrapped up in tax or accounting jargon, that describes exactly how distributions will be made. Pay special attention to how they will be made. Also when there are IRR hurdles that have to be accomplished, how those will be calculated?

9. Communication

    The ninth key thing to look for in these deals is communication. Is the sponsor communicating as they said they would? Are they being responsive to post-close inquiries that you or other members may have? Are they getting documents and reports to you in a timely fashion? Do they demonstrate that they're on top of the project and that they run a tight operation?

10. Exit strategies

The tenth key thing to look for is exit strategies. Has the sponsor outlined a series of exit strategies in their documentation? Have they shown you some scenarios whereby they sell or refinance the property? Or if the market goes down, can they hold it over the long term? Will they sell the whole property? Or will they parcel it off?

Has the sponsor provided the best, worst, and most likely scenarios? Having a well-formulated, detailed description of various exit strategies shows a sophisticated, competent sponsor.

Seven red flags

When you take a look at contracts, there are potential red flags that may warn you off the deal. Here are seven of the most important ones.

1. Aiming too high

The first red flag is the unintended consequences of aiming too high. Remarkably, some projects were still under water a decade or more after the last recession. It's incredible to imagine that ten years later, there were still projects that had not emerged from the problems of that time. A practical example of this is broken condo deals with homeowner associations that were triggered during the Great Recession but where entitlement issues have slowed down efforts to address the problems.

Be sure to understand where the entitlements are on a project. How certain is the sponsor that they're going to get them? If they don't have them, that's definitely a red flag. It's not a trivial thing to get entitlements.

Also, keep an eye out for loan expirations. What are the durations of the loans that the sponsor's taken down? Considering the business plan that has been proposed, are they of reasonable length? What's more, look for scale transformations. If you have a Class C property in a C area, but the sponsor plans to build a Class A property in that area, the proposed project might not be a good idea. It could end up being overbuilt for the area. If a deal seems too good to be true, it probably is.

2. Cycle-appropriate IRR

Keep in mind that institutional investors shy away from super-high IRR product at later stages of the cycle. If you're looking at deals that are at the highest end of the projected return scale, then see that as a red flag. A 25 percent IRR with a three- to five-year life cycle on the deal might be great in the early days of recovery, but it might be aggressive during the hypersupply and recession stages of the cycle.

Whatever stage of the cycle you are investing in, look at the projected returns relative to other prudent returns in the market. If the projected returns are out of whack with what other people are doing or with what institutional investors are looking at, it's probably overly ambitious. In these cases, it's probably best to stay away from the deal.

3. Timeline issues

Take a look at a sponsor's projections. If you're looking at three to five years out, is it realistic that they will hit their projected results? Do they acknowledge that the coming market cycle will be the same as *you* think it's going to be? If you think the market's going down, is the sponsor also predicting the same kind of trend? If you think the market's going up, then you also want to see the sponsor projecting the same trend.

As long as the timeline projections are consistent with your view of where we are in the market, then at least you're aligned in that regard with the sponsor, going in. If you are not, that's a red flag. As you make this decision, think of the implications in the contract for your rights and remedies in case things don't go as planned.

4. Voting rights

    The fourth red flag is voting rights. If you see limited or nonexistent voting rights, that is an enormous red flag. In some contracts, for example, I've seen an expressed prohibition on knowing who the other investors are.

    No matter what voting rights you have, if you cannot call a vote and if you don't know who to ask for a vote, then you are in deep water. In that situation, you have very few rights; the deal is a semi-dictatorship in which you won't be able to vote on anything. If you want to give your money to a semi-dictator, be my guest. For me, that's a red flag. Definitely keep an eye out for voting rights.

5. The manager's reporting obligations

    The fifth red flag is the reporting obligations of the manager. If there are no, or very limited, reporting requirements, then that's problematic. You may have heard that quarterly reports are mandatory. In some cases, that might be state-specific (again, I'm not a lawyer). At the same time, I have seen many contracts where they offer to report only once a year. The rest of the year, there is very, very limited or highly restricted access to the books and records.

6. No notice must be given to investors

    If a deal doesn't require notice be given to the investors, under certain circumstance that's also a red flag. For example, the sponsor could put new debt on the deal but isn't obligated by the contract to notify you. Or new investors could be brought into the deal by the sponsor without your knowledge.

    You don't want to wake up tomorrow and find out that layers of debt have been put on the project in positions of priority payment ahead of you, or that new investors have been brought in, whose rights and terms are all superior to yours. If that happens, you may have just been squeezed out of the deal through dilution. Instead, you want to know what's going on. Notice must be given ahead of those things. And if you don't think that what they're planning to do is a good idea, you need to have some remedies.

7. Deal complexity

    The seventh red flag is deal complexity. If you can't explain the nuances of the deal to somebody else, don't invest. If there's a section in there you simply can't explain to somebody, as though you were pitching the deal to them, you probably shouldn't enter the deal.

Eight ways to protect yourself

Finally, here are some ways to protect yourself as you consider investing in a project.

1. Outside advice

    The first thing is to seek outside advice. Get a mentor or find an expert in the asset type that you're considering. Ask them cast their eyes over a deal before you decide to jump in.

2. Look at the bank

    The second way of protecting yourself is to think about the bank that's involved. Look at the caliber of the lending institute. Take a look at the loan application and the package to see what it includes and ask the sponsor what they sent to the lender. Find out how diligent the bank was in examining this sponsor and the deal before they went in. If it's a credible bank that is high up on the 'stress test' status at the Federal Reserve, then they've likely done considerable due diligence to pass this deal—due diligence that you would otherwise do yourself.[5]

The bottom line is that having a good lender on a deal can be a good sign. It's another way to get some assurance that a deal has credibility.

3. Invest in assets you understand

    The third way to protect yourself is to invest in an asset type in which you have some experience, or at least some intuitive understanding. It's probably one of the reasons that multifamily and single-family homes are such a popular asset type. After all, everybody lives somewhere. Invest in something that you understand, and look to sponsors who provide deep and detailed educational content on their website and in their newsletters. This should not only explain the business they are in, but also provide a coherent explanation of their strategy.

4. Monitor the deal post close

    The fourth way to protect yourself is monitor the deal post close. Don't go to sleep once the deal is closed. Visit the site. Read the reports. Ask for the key documents once in a while to ensure that the sponsor is responsive. You don't need to be annoying, but you can occasionally ask for a document. Make sure that the sponsor is on their toes and that things are in order. Also, check the finances for the project to see that the sponsor is on budget.

    In general, you want to stay ahead of the curve to make sure things aren't going south. Since you don't want to suddenly discover other investors diluting you, stay tuned in to the budget and monitor the sponsor's expenditures.

5. Debt versus capital

    The fifth way to protect yourself is to stay aware of the debt versus capital balance in the deal. Make sure the sponsors bring in more capital as debt from investors before a third-party lender is called or they bring in new investors who might dilute you. If you can get it, you want to see debt coming in before capital that would supersede and potentially dilute you.

6. Right of access to the books

    The sixth way to protect yourself is to make sure that you have a right of access to the books. In many cases, this means reasonable notice—anything from 48 hours to two weeks. Again, you don't want to be a pain, but you certainly want to have access. Once in a while, use this to check in with the manager to make sure that they are responsive. This is similar to monitoring the deal. This will enable you to take action if the sponsor becomes nonresponsive or noncommunicative.

7. Communication

    Right of access takes us to the seventh way to protect yourself: communication. Make sure to continue communicating with the sponsor post close, to ensure they remain communicative with you. If possible, stay in touch with the other investors as well. Find out who they are, share ideas, learn from them, and monitor the project together.

8. Read everything

    The final way to protect yourself is the blindingly obvious but often disregarded: Read everything. Read every single document. Take a holistic approach and ensure that everything in the documents is in sync not only within the documents themselves but also with how you want a deal to be structured.

Personally, I read everything, but when I first look at a deal, I take shortcuts as an initial screen. To expedite the process, I scan for key concepts. If all of these key concepts look like they're okay, *then* I read everything. For me, scanning means running word searches and then looking at the context where the sponsor uses these words.

In a long PDF, you can search for terms like:

- risks
- sources and uses
- fees
- distributions
- liquidity
- dispute resolution
- voting rights

By scanning the document, you quickly get to the core of the most important matters for you. If you discover that the contract uses a different keyword, run a search for that. Especially if the term is a trigger driver for you.

Now it is time for you to go out and have some fun! Start digging into the contracts you see out there. Take a good look at some online deals right now. Go and pull down some of the private placement memorandums and the offering documents and run through them. Run word searches to fast-forward to key concepts. Read them in context. Check out the terms.

You've never had such access before. It's only since the advent of crowdfunding that you can do this—so take advantage of the opportunity. The more you see, the more proficient you will become in identifying deals that you simply don't want to get into. Conversely, you'll become better at identifying deals where the sponsor has put forward a fair, reasonable, good-faith contract that you can live with. The more you see, the easier it is for you to identify a deal with an acceptable set of offering documents.

Notes

1 I am not a lawyer and this chapter is not meant to replace your need to hire one to check your contract and advise you on anything you might be considering signing.
2 Actually, I've seen more downturns than just two. I was investing very heavily in real estate across Japan during that country's real estate meltdown during the 1990s, and I also survived the dot-com bust of the early 2000s, though the impact on real estate during that downturn was less pronounced than during others.
3 My general rule for all things related to tax, here and throughout this book and in everything I write, is the AAA rule: If it pertains to tax, Ask An Accountant. The AAA rule applies equally to any questions of law: Ask An Attorney.
4 The contracts in any real estate transaction are collectively described as the 'offering documents.' These can include a private placement memorandum, the operating agreement, and the subscription agreement. Yet there could be other contracts in there as well. They might have different names. Specifically, we're focusing on the entire corpus of these offering documents, the mortar which binds the layers together.
5 You can find the bank ratings scale at the Federal Reserve website: federalreserve.gov

10 Conclusion

Conquering real estate crowdfunding

The philosopher and scholar Erasmus, one of the greatest minds of the Renaissance period, once wrote: "In the land of the blind, the one-eyed man is king." The preceding pages partially opened your eyes, and you're on your way to becoming ruler of all you survey. But you won't be able to truly see until you get out into the world and start finding and getting involved with crowdfunding deals in real life. Until then, all of this wisdom is just a mental exercise. In our final time together, I want to offer some parting words, some of caution, some of advice, but most of all, words of encouragement to get you started on your journey.

Proceed with caution

Before I get you pumped up, I want to offer a word of caution. My goal with this book is not to sell you an unattainable dream. As a real estate investment and finance educator at universities and online, I firmly believe that my students' success is my success. I want you to win. Real estate crowdfunding has a clear path to that success, with tremendous upside as an investment class; but there is one crucial caveat: The same principles that apply to prudent investing in other sectors are just as relevant, if not more relevant, in the real estate crowdfunding space.

Any investment advisor worth their salt will tell you that diversification is the first step to a healthy portfolio. What this means for you, practically, is that you should not and cannot keep too much of your wealth/retirement savings in crowdfunded real estate deals. Everyone's risk tolerance is different—a middle-aged worker creeping up to retirement will want safer routes, while a recently graduated go-getter may have a much higher appetite for risk. Before deploying any capital in crowdfunded syndications, be sure you are approaching your investment as part of a well-thought-out, diversified strategy—you can reap the rewards offered by crowdfunding without putting you or your family in financial danger.

Now let's get to the good stuff.

The good stuff (why crowdfunding works)

After a book filled with fractions, formulas, and scenarios, I think it would help to remind you why you were drawn to this asset class in the first place.

1. Since 2000, real estate has substantially outperformed the stock market.

As an investor, you're well aware of the robust performance of the stock market over the past ten years or so (at time of writing, of course). You can't turn on CNBC or Bloomberg or open up a copy of *The Wall Street Journal* or *Barron's* without witnessing or reading headline after headline trumpeting another record day for the equities markets. So unless you knew better, you would assume that the stock market was the most lucrative place to put your money in the preceding years; but that would be a bad assumption.

Real estate assets have outperformed the stock market since 2000, earning more than 10 percent annually, while stocks barely cracked 5 percent.[1] This does not mean that real estate will outperform stocks in the future; nor does it mean that every real estate project will provide a better ROI than every equity investment. But it shows you that real estate is a powerful investment class with real potential for generating wealth, and one worth considering for your portfolio.

2. Crowdfunding is far more accessible than purchasing and/or managing commercial real estate assets on your own.

The commercial real estate industry could be producing 20 percent or even 30 percent returns, but it wouldn't mean a thing to most investors, many of whom lack the capital to invest tens or even hundreds of millions of dollars into commercial real estate projects. The fractional ownership model in the crowdfunding space allows investors to get in on the action without having to place a big bet—you can invest for thousands of dollars, compared to hundreds of thousands or millions for direct ownership investments. This ability to buy in for a small piece of many projects, rather than a single large investment in a commercial property, brings me to my next point.

3. Crowdfunding makes diversifying your real estate portfolio easy.

Real estate is not exactly the easiest asset class to diversify. In fact, many Americans are actually holding an outsized portion of their retirement savings in real estate in the form of their home, with a little more than 64 percent of Americans owning their own homes. Now, there is a discussion to be had about whether or not one should see one's primary residence as an investment, but that attitude does exist. Plenty of retirees hold a not insubstantial portion of their savings in their homes, particularly in high-cost-of-living regions like New York and San Francisco.

With crowdfunding, you can gain exposure to real estate assets in the type and sectors of your choice, whether that is self-storage, multifamily, retail, or any other commercial real estate operation. More importantly, you can invest amounts significantly lower than those you might invest in your home. A diverse portfolio is a safer portfolio, and crowdfunding allows investors in every stage of life to benefit from adding assets other than traditional stocks and bonds to their retirement savings.

4. Crowdfunded real estate may offer you a hedge against failure in the broader economy.

As I am finishing up writing this book, we are well into our 11th year of a bull market. Some commentators, financial journalists, and media pundits seem to believe that the party can go on forever. Unfortunately, from personal experience and even just a tertiary glimpse at historical trends, I know this simply cannot be true, no matter how much we wish it to be so. Business cycles and market contractions are inevitable—we have not yet managed to eradicate the fundamental laws that govern our economic system, no matter which party is in power or how pro-business they are. For those of us that remember the 2008 crash, the memories of those who lost everything when the music stopped remain fresh.

The beauty of crowdfunded real estate syndications is the diversity of potential investments, especially those with countercyclical resistance to economic contractions or recessions. Property types like self-storage facilities, multifamily apartments, and even some types of retail, office, and industrial properties thrive when things turn sour for the broader economy. Oftentimes, investors looking for a hedge choose to invest in precious metals like gold and silver, or bonds; however, these asset classes cannot match the returns and long-term consistency of commercial real estate.

Remember Warren Buffett's rule: Invest in things that create value. Gold, silver, and stocks and bonds are all limited in what they can do, because much of their value is speculative. But real estate has real value and can pull in a regular income for investors, which is what you want in an investment—not something that has value because it is shiny or because people are flocking to the next big thing, as is the case with many cryptocurrencies.

Final thoughts

The best way to succeed, in life and in business, is to emulate those who have already accomplished the goals you have set for yourself. Until very recently, large-scale commercial real estate deals, like those found on crowdfunding platforms, were the exclusive province of the elites, the 1 percent, not the plebeians. Many of the wealthiest people in our society became that way through real estate, and almost every single high-net-worth individual in the country owns significant real estate assets, whether it be land or income-generating properties, or through investments in mortgage lenders and real estate service providers.

Now that you, as an individual investor, have the ability to build wealth in a similar way, don't squander the opportunity. Use the strategies and game plan laid out in this book to write your own success story—one where you **conquer the world** using the **power of real estate crowdfunding**.

Note

1 Real estate outperformed the stock market by approximately 2:1, returning 10.71 percent annually compared with the S&P 500 Index's 5.43 percent annual total return.

Addendum 1: A starter guide on preferred equity[1]

Preferred equity is one of those financial concepts that a lot of investors believe they understand, and yet each may have a different concept in mind, which can lead to confusion and misunderstanding.

This article provides guidance on the definition of preferred equity, and how you can tell the difference between the different types.

Simply stated, there are two types of capital that are used in any real estate transaction: debt and equity.

The main difference between these two types of capital is that debt will only be repaid with interest, whereas equity is commonly paid a type of interest (usually called a 'preferred return') plus, importantly, a share of the profits.

Each of these categories of finance can be broken down into different subcategories.

For example, there are different kinds of debt: senior debt, most often provided by banks; second-position debt, which is subordinate to senior debt; and even third- and fourth-position debt, and so on, each subordinate to each other sequentially.

All types of debt are considered loans to the property and are typically secured by recorded liens on the property.

Liens against the underlying real property are put in place so that should the borrower stop paying as agreed to any lender, there is a formal record of the obligation.

Using the foreclosure process, lenders can institute recovery efforts, potentially culminating in their taking possession of a property away from a borrower.

Mezzanine debt

Mezzanine debt financing comes last after other types of debt, is always subordinate to any other recorded debt, but is ahead of all equity.

However, mezzanine debt does not record a lien against the property itself as collateral.

Instead mezzanine debt secures its position with a claim against the equity in the deal via a UCC [Uniform Commercial Code] financing statement filed against the partnership interest.

Despite not securing its position against an interest in the underlying real property, mezzanine financing is considered a loan to the project and mezzanine holders are considered to be lenders, which affords them different recovery rights than equity holders.

The capital stack

Capital stack

At the bottom of the stack (see Figure 8.6 in Chapter 8) is the most stable of all the capital—the foundational capital upon which all other funding is placed: the bank debt, with a first-position mortgage, also known as senior debt.

On top of that might be another layer of debt, the second-position debt.

Then follows the mezzanine debt, then the equity, which might be split between preferred and common equity.

Generally, second-position debt will earn an interest rate higher than that in first position because it bears higher risk.

Mezzanine debt is so called because it sits in between two different types of capital and is the last layer of financing before the equity layer in the capital stack.

Preferred equity

During the financial crisis of 2007–2008, banks discovered challenges foreclosing on properties that had a layer of mezzanine debt on them.

Mezzanine holders, acting in their capacity as lenders to projects, were able to exercise outsized influence on the control of the real property, and therefore banks now include in their loan documents prohibitions against adding extra layers of debt, including mezzanine debt, on top of the senior position.

To fill the gap that the absence of mezzanine debt left, preferred equity is increasingly used in its place.

Like mezzanine debt, preferred equity is subordinate to all debt and superior to common equity.

Preferred equity does not receive a share of the profits of the transaction, and yet, as its name suggests, preferred equity is not debt and investors in preferred equity are not treated like lenders.

But if preferred equity investors are not considered lenders to the project—a key contrast to mezzanine debt—and don't share in the profits, as typical equity investors might expect, what are they?

The mechanism by which investors can secure their interests that differentiates preferred equity from mezzanine debt is as follows:

Preferred equity holders, as members of the LLC [limited liability company] controlling the real estate, can in the event of default take over control of the LLC.

In the case of preferred equity, recourse is strictly a function of the structure of the operating entity and the relationship between the various parties.

And because preferred equity holders are simply a different class of shareholder in the company that is operating the real estate project, they do not represent another layer of secured debt and consequently pose reduced compliance issues with lenders.

Mezzanine debt holders are not members of the LLC, but their recourse is to take possession of the LLC in its entirety in the event of default, through a UCC foreclosure.

This contrasts with a debt lender with a lien against the property, who would take over ownership of the real property itself—preferred equity and mezzanine debt investors take over ownership of the partnership that owns the property.

While the intent of preferred equity is to fill whatever gap was left by mezzanine debt, there is a broad array of different rights and remedies that can be built into an operating agreement.

Consequently, sophisticated investors should take into consideration what these rights and remedies are when comparing different preferred equity investments—no two 'preferred equity' structures are the same and, in fact, can be very different.

Soft versus hard pay

One way in which preferred equity positions can be very different from each other is the manner in which they are paid: soft or hard.

Hard pay functions more like a debt instrument, where if there's a nonpayment, there are some punitive remedies.

The operating agreement will say that the operating entity has to make a certain payment on the first of every month and pay off the entire loan within a certain period of time—let's say 3 years.

If the entity, controlled by the sponsor, doesn't make any of those payments, pure 'hard pay' remedies that could be enforced could include wiping out all the subordinate equity and taking over control of the partnership interest.

In this sense it is pretty much identical to mezzanine debt.

And it is precisely for this reason that banks don't like hard-pay preferred equity positions.

The bank is lending to the sponsor and does not like subordinate positions to have the right to take over the operating entity they are lending to.

Soft-pay terms, more acceptable to lenders, just require the sponsor, or the operating entity, to make payments when there is sufficient cash flow to be able to make payments, and if there is a failure to pay, there may or may not be remedies the preferred equity holders can employ.

Furthermore, while preferred equity is paid before common equity, there may be 'hard pay' terms that set aside a preferred interest reserve.

The preferred equity holders will be paid out of this reserve, even though common equity holders may receive the first distributions of income from cash flow.

Risks to unsophisticated investors

This combination of different interests and options can create additional risks for unsophisticated investors.

If a sponsor puts together a deal that works in their best interests, like a soft-pay preferred equity deal, they may be able to put this in front of a crowd that doesn't understand whether the deal is good or bad.

In essence, it could allow the sponsor to have a free option with the preferred equity money.

Consequently, there are vast differences between types of preferred equity.

Investors should be keenly aware that their rights can be materially affected in one kind of preferred equity versus another and should be conscious of what their rights are before making an investment.

Note

1 This article first appeared on my website and can be found at: https://gowercrowd.com/learn/preferred-equity

Addendum 2: The impact of crowdfunding on real estate waterfall structures[1]

Waterfall calculations are used by real estate developers to calculate distributions to investors during the life cycle of a project, and there have been some notable shifts in how they are structured since the emergence of real estate crowdfunding—with significant implications.

One trend is the way investors are driving how sponsors underwrite and present their deals, the tail wagging the proverbial dog, and another pertains to the co-investment a sponsor puts into a project to align interests with investors. As you will see in this article, when these two factors converge, pricing is driven upward by hypersupply of liquidity, risk increases, and we enter the last throes of the cycle.

Show me the IRR!

In the recovery phase of real estate cycles, those periods following downturns and recessions, investors are enticed back into real estate with the promise of higher returns to offset the perceived excess of risk. Furthermore, the opportunity to invest in 'distressed' deals that are available at discounts to peak values motivates investors who expect higher returns founded upon the bargain pricing they see. As the economy recovers, opportunistic investments diminish in frequency, and expectations are for returns to gradually reduce. Even so, sponsors continue to attract investors because margins become tighter for everyone as the market reaches equilibrium.

The real estate crowdfunding paradox is that the driving force behind investor decisions to invest in individual deals is the IRR. Institutional and other sophisticated investors understand that returns diminish as the cycle prolongs and set expectations accordingly; crowd investors as a group do not—yet.

Retail investors, those coming into real estate investing through crowdfunding platforms, are signaling to sponsors that they think differently. In recent investor surveys conducted by some of the major crowdfunding sites, investors reported being primarily driven by IRR—the higher the better. Reinforcing this quantitative result, crowdfunding websites anecdotally report that projects with the highest projected IRRs tend to have significantly higher webinar attendances, suggesting investors are chasing maximum returns even if disproportionate to what would normally be expected.

This is counterintuitive to the stage of the cycle we are currently in. Lower, not higher, IRRs signal the natural trajectory of prudent underwriting, and yet sponsors know that to attract crowd investors, they must offer projects with only the highest relative IRRs or lose investors to someone else's deal.

Consequently, there is a rush to compete with headline IRR numbers. Crowdfunding sponsors know that the sponsor with the biggest IRR wins. The danger of this is that it can motivate sponsors to 'stretch' underwriting assumptions to deliver higher headline IRR numbers to attract the unwary investor. This leads to an unnatural inversion of the IRR curve, where competition for investor capital drives up projected IRRs during the latter stage of the cycle.

This is a sure sign of a market that is overheating.

The law of diminishing co-investment

Another anomaly as the market goes into its final upward lurch is related to the co-investment sponsors contribute to projects. The co-invest is the risk capital a sponsor personally contributes to a project as part of their alignment of interest with investors.

Investor Management Services (IMS), the leading investment management software company in the industry, has noticed that the amount sponsors co-invest in projects varies with the market cycle. In the early part of a recovery when perceived risk is higher and yet actual risk is lower, sponsors are expected to contribute more to ensure a closer alignment of interest with investors.

As risk climbs inexorably towards the end of the cycle as pricing reaches peak levels and supply moves to hypersupply conditions, perceived risk diminishes, and yet IMS observes that sponsors are required to put in less and less of their own money.

Bubble blooming

As these two factors, irrational IRR inflation and reduced sponsor co-investment, converge, expectations rise, interests become misaligned between sponsor and investor, risks expand and pricing increases—classic indicators of the final stages of a rising bubble.

To alleviate the risks of getting caught up in these dynamics, investors can pay attention to the credentials of a sponsor, particularly their experiences during prior recessions, and resist the temptation to seek out only the project with the highest, shiniest IRR projections. The best projects may well be those with lower IRRs, because their sponsors are taking a cautious approach as we near the end of the cycle and could be those most likely to survive the inevitable downturn.

Note

1 You can find this article at: https://gowercrowd.com/real-estate-syndication/impact-of-crowdfunding-on-real-estate-waterfall-structures There are several other articles discussing advanced waterfall concepts on my website, plus you can gain access to the white paper I wrote at: https://gowercrowd.com/learn/whitepaper

Bibliography

Building Owners and Managers Association International [website] www.boma.org/

Federal Reserve [website] https://federalreserve.gov

Gower, A. (2019) "A primer on preferred equity," GowerCrowd, https://gowercrowd.com/learn/preferred-equity

Gower, A. (Host). (2019) "Ben Miller, Co-Founder, CEO Fundrise (204)" [audio podcast episode] in *The Real Estate Crowdfunding Show: Syndication in the Digital Age*, GowerCrowd, https://gowercrowd.com/podcast/204-ben-miller-co-founder-ceo-fundrise

Gower, A. (Host). (2019) "Brew Johnson, PeerStreet (206)" [audio podcast episode] in The Real Estate Crowdfunding Show: Syndication in the Digital Age, GowerCrowd, https://gowercrowd.com/podcast/206-brew-johnson-peerstreet

Gower, A. (Host). (2019) "Buying into a bubble (104)" [audio podcast episode] in *The Real Estate Crowdfunding Show: Syndication in the Digital Age*, GowerCrowd, https://gowercrowd.com/podcast/004-buying-into-a-bubble

Gower, A. (Host). (2019) "Chris Loeffler, Founder/CEO Caliber Companies (216)" [audio podcast episode] in The Real Estate Crowdfunding Show: Syndication in the Digital Age, GowerCrowd, https://gowercrowd.com/podcast/216-chris-loeffler-founder-ceo-caliber-companies

Gower, A. (Host). (2019) "Chris Rawley, Founder/CEO Harvest Returns (337)" [video podcast episode] in The Real Estate Crowdfunding Show: Syndication in the Digital Age, GowerCrowd, https://gowercrowd.com/podcast/chris-rawley-founder-ceo-harvest-returns

Gower, A. (Host). (2019) "Chuck Schreiber, CEO/Co-Founder KBS (222)" [audio podcast episode] in *The Real Estate Crowdfunding Show: Syndication in the Digital Age*, GowerCrowd, https://gowercrowd.com/podcast/222-chuck-schreiber-ceo-co-founder-kbs

Gower, A. (Host). (2019) "Eve Picker, Small Change (209)" [audio podcast episode] in The Real Estate Crowdfunding Show: Syndication in the Digital Age, GowerCrowd, https://gowercrowd.com/podcast/209-eve-picker-smallchange

Gower, A. (Host). (2019) "Historic preservation, The Penn Central case (109)" [audio podcast episode] in The Real Estate Crowdfunding Show: Syndication in the Digital Age, GowerCrowd, https://gowercrowd.com/podcast/009-historic-preservation-the-penn-central-case

Gower, A. (Host). (2019) "Larry Feldman, CEO Feldman Equities (332)" [video podcast episode] in The Real Estate Crowdfunding Show: Syndication in the Digital Age, GowerCrowd, https://gowercrowd.com/podcast/larry-feldman-ceo-feldman-equities

Gower, A. (2019) *Leaders of the Crowd* (New York, Palgrave MacMillan). Early syndication chapter available at: https://leadersofthecrowd.com/

Gower, A. (Host). (2019) "Lew Feldman, CEO Heritage Capital (223)" [audio podcast episode] in The Real Estate Crowdfunding Show: Syndication in the Digital Age, GowerCrowd, https://gowercrowd.com/podcast/223-lew-feldman-ceo-heritage-capital-ventures

Gower, A. (Host). (2019) "Max Sharkansky, Managing Partner, Trion Properties (219)" [audio podcast episode] in The Real Estate Crowdfunding Show: Syndication in the Digital Age, GowerCrowd, https://gowercrowd.com/podcast/219-max-sharkansky-managing-partner-trion-properties

Gower, A. (Host). (2019) "Nav Athwal, CEO/Founder, RealtyShares (202)" [audio podcast episode] in The Real Estate Crowdfunding Show: Syndication in the Digital Age, GowerCrowd, https://gowercrowd.com/podcast/202-nav-athwal-ceo-founder-realtyshares

Gower, A. (Host). (2019) "Real Estate Cycle Monitor Report (110)" [audio podcast episode] in The Real Estate Crowdfunding Show: Syndication in the Digital Age, GowerCrowd, https://gowercrowd.com/podcast/010-real-estate-cycle-monitor-report-background

Gower, A. (Host). (2019) "Real estate value investing (118)" [audio podcast episode] in The Real Estate Crowdfunding Show: Syndication in the Digital Age, GowerCrowd, https://gowercrowd.com/podcast/018-real-estate-value-investing

Gower, A. (Host). (2019) "Robert T. O'Brien, Vice Chairman Deloitte (108)" [audio podcast episode] in The Real Estate Crowdfunding Show: Syndication in the Digital Age, GowerCrowd, https://gowercrowd.com/podcast/robert-t-obrien-vice-chairman-deloitte

Gower, A. (Host). (2019) "Steven Kaufman, Zeus Crowdfunding (235)" [audio podcast episode] in The Real Estate Crowdfunding Show: Syndication in the Digital Age, GowerCrowd, https://gowercrowd.com/podcast/235-steven-kaufman-zeus-crowdfunding

Gower, A. (2019) "The impact of crowdfunding on real estate waterfall structures," GowerCrowd, https://gowercrowd.com/learn/impact-of-crowdfunding-on-real-estate-waterfall-structures

Gower, A. (2019) "What is the optimal investment period for real estate," GowerCrowd, https://gowercrowd.com/learn/what-is-the-optimal-investment-period-for-real-estate

Gower, A. (2019) "White paper on waterfall structures," GowerCrowd, https://gowercrowd.com/learn/whitepaper

Gower, A. (2019) "Why RealtyShares failed," GowerCrowd, https://gowercrowd.com/podcast/239-nav-athwal-realtyshares-special

Gower, A. (Host). (2019) "Will the crowd fund the next generation of retail (201)" [audio podcast episode] in The Real Estate Crowdfunding Show: Syndication in the Digital Age, GowerCrowd, https://gowercrowd.com/podcast/201-will-the-crowd-fund-the-next-generation-of-retail

Ringer, Robert J. (1974) *Winning through Intimidation* (Los Angeles, Fawcett Crest Books).

Shopping for the Truth [website] https://partners.wsj.com/icsc/shopping-for-the-truth/maximizing-the-human-experience

Index